Grade 8 · Unit 1

Inspire Science

Change Over Time

Phenomenon: The Caterpillar

Caterpillars, like the one on the cover, have soft bodies with bright colors that can, in some cases, indicate that it is poisonous. The caterpillar has learned in many ways to defend itself including, mimicking a predator to protect itself from being eaten.

Some caterpillars have 12 eyes.

FRONT COVER: (t)Darlyne A. Murawski/National Geographic/Getty Images, (b)Damsea/ Shutterstock. **BACK COVER:** Darlyne A. Murawski/National Geographic/Getty Images.

mheducation.com/prek-12

Send all inquiries to:
McGraw-Hill Education
STEM Learning Solutions Center
8787 Orion Place
Columbus, OH 43240

ISBN: 978-0-07-687489-7
MHID: 0-07-687489-3

Printed in the United States of America.

1 2 3 4 5 6 QVS 22 21 20 19 18

McGraw-Hill is committed to providing instructional materials in Science, Technology, Engineering, and Mathematics (STEM) that give all students a solid foundation, one that prepares them for college and careers in the 21st century.

Welcome to

Inspire Science

Explore Our Phenomenal World

Learning begins with curiosity. Inspire Science is designed to spark your interest and empower you to ask more questions, think more critically, and generate innovative ideas.

Start exploring now!

Inspire Curiosity • Inspire Investigation • Inspire Innovation

Authors, Contributors, and Partners

Program Authors

Alton L. Biggs
Biggs Educational Consulting
Commerce, TX

Ralph M. Feather, Jr., PhD
Professor of Educational Studies and Secondary Education
Bloomsburg University
Bloomsburg, PA

Douglas Fisher, PhD
Professor of Teacher Education
San Diego State University
San Diego, CA

Page Keeley, MEd
Author, Consultant, Inventor of Page Keeley Science Probes
Maine Mathematics and Science Alliance
Augusta, ME

Michael Manga, PhD
Professor
University of California, Berkeley
Berkeley, CA

Edward P. Ortleb
Science/Safety Consultant
St. Louis, MO

Dinah Zike, MEd
Author, Consultant, Inventor of Foldables®
Dinah Zike Academy, Dinah-Might Adventures, LP
San Antonio, TX

Advisors

Phil Lafontaine
NGSS Education Consultant
Folsom, CA

Donna Markey
NBCT, Vista Unified School District
Vista, CA

Julie Olson
NGSS Consultant
Mitchell Senior High/Second Chance High School
Mitchell, SD

Content Consultants

Chris Anderson
STEM Coach and Engineering Consultant
Cinnaminson, NJ

Emily Miller
EL Consultant
Madison, WI

Key Partners

American Museum of Natural History

The American Museum of Natural History is one of the world's preeminent scientific and cultural institutions. Founded in 1869, the Museum has advanced its global mission to discover, interpret, and disseminate information about human cultures, the natural world, and the universe through a wide-ranging program of scientific research, education, and exhibition.

SpongeLab Interactives

SpongeLab Interactives is a learning technology company that inspires learning and engagement by creating gamified environments that encourage students to interact with digital learning experiences. Students participate in inquiry activities and problem-solving to explore a variety of topics through the use of games, interactives, and video while teachers take advantage of formative, summative, or performance-based assessment information that is gathered through the learning management system.

Phet Interactive Simulations

The PhET Interactive Simulations project at the University of Colorado Boulder provides teachers and students with interactive science and math simulations. Based on extensive education research, PhET sims engage students through an intuitive, game-like environment where students learn through exploration and discovery.

Table of Contents
Change Over Time

Module 1 Geologic Time

Module 2 Natural Selection and Adaptations

Module 3 Evidence of Evolution

Change Over Time

Geologic Time

ENCOUNTER THE PHENOMENON

How can rocks tell the story of Earth's long history?

Hutton's Unconformity

GO ONLINE
Watch the video *Hutton's Unconformity* to see this phenomenon in action.

Collaborate With your class, develop a list of questions that you could investigate to find out how scientists use clues from rocks to organize Earth's history. Record your questions below.

Lesson 1
Analyzing the Rock and Fossil Records

Lesson 2
Building a Time Line

STEM Module Project Launch
Science Challenge

History of Rock

Maria is in Ms. Monroe's science class. She and her partner, Sam, are studying a map of the last ice age. Maria looks thoughtful. She raises her hand.

"Yes, Maria?" Ms. Monroe asks.

"How did scientists come up with this map?" Maria asks. "There aren't any written records. What's their evidence?"

"Funny, I was going to ask you the same question," Ms. Monroe says with a smile. "Your homework assignment is to explain how rocks and fossils provide evidence about major events in Earth's history, like the last ice age, and how this evidence is used to organize the geologic time scale. And be creative! Think of a fun way to present your explanation."

Maria and Sam look at one another. That sounds like a lot of work! They'll need your help to construct a video blog that explains how evidence from rock strata is used to organize Earth's vast geologic past.

Start Thinking About It

What kind of evidence can you use to construct a scientific explanation? Discuss your ideas with the class.

STEM Module Project
Planning and Completing the Science Challenge
How will you meet this goal? The concepts you will learn throughout this module will help you plan and complete the Science Challenge. Just follow the prompts at the end of each lesson!

PAGE KEELEY SCIENCE PROBES

LESSON 1 LAUNCH

How old is Earth?

Haley wonders which time units she should use to describe the age of Earth. Circle what you think are the best units to describe Earth's age.

Hundreds of years

Thousands of years

Millions of years

Billions of years

Trillions of years

Explain your choice of time units. Include what you think is a reasonable estimate of the age of Earth.

____

____

____

____

____

____

____

You will revisit your response to the Science Probe at the end of the lesson.

LESSON 1

Analyzing the Rock and Fossil Records

ENCOUNTER THE PHENOMENON | How can the sequence of past geologic events be determined?

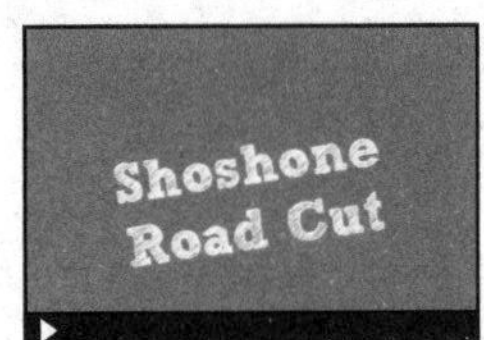

GO ONLINE

Explore the *Shoshone Road Cut* to see this phenomenon in action.

What did you notice as you explored the photo online? Try to identify distinct parts of the image and then focus on one part at a time. Record your observations in the space provided. How might you determine which layers and features are older than others? With your partner, discuss the clues you might use to determine the sequence of the past geologic events that created this landscape.

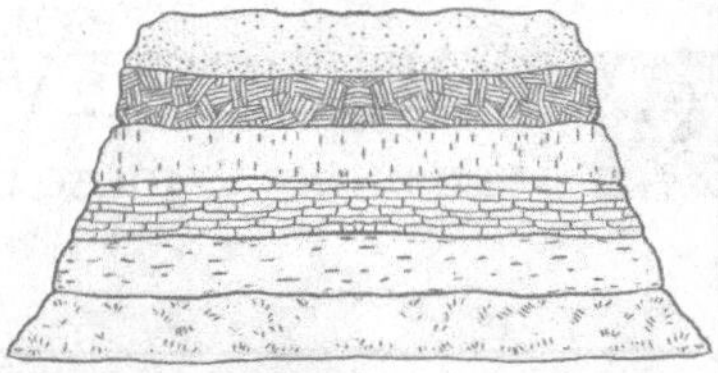

EXPLAIN THE PHENOMENON

Are you starting to get some ideas about the clues geologists use that enable them to read rocks layers like pages in a history book? Use your observations about the phenomenon to make a claim about how the order of geologic events can be determined.

CLAIM

The sequence of past geologic events can be determined by...

COLLECT EVIDENCE as you work through the lesson.

Then return to these pages to record your evidence.

EVIDENCE

A. What evidence have you discovered to explain how the rock record provides relative ages?

MORE EVIDENCE

B. What evidence have you discovered to explain how the fossil record provides relative ages?

When you are finished with the lesson, review your evidence. If necessary, based on the evidence, revise your claim.

REVISED CLAIM

Scientists can determine the sequence of past geologic events by...

Finally, explain your reasoning for how and why your evidence supports your claim.

REASONING

The evidence I collected supports my claim because...

What is the basis for understanding Earth's past?

Early ideas about Earth's age and geologic history were usually placed in the context of time spans that a person could understand relative to his or her own life. This changed as people began to explore Earth and Earth processes in scientific ways. Today, scientists know that Earth's history stretches back 4.6 billion years. When did this change of thought occur, and how?

INVESTIGATION

The Present Is the Key to the Past

Compare the images of erosion below.

1. Do you think the processes that form and shape the small stream bed are similar to those that form and shape the Grand Canyon? Why or why not?

2. How long do you think it would take to create a canyon as deep as the Grand Canyon? Explain your reasoning.

Want more information?
Go online to read more about how scientists analyze the rock and fossil records.

FOLDABLES®
Go to the Foldables® library to make a Foldable® that will help you take notes while reading this lesson.

Principle of Uniformitarianism Before the late 1700s, most people thought that Earth was only a few thousand years old. James Hutton rejected this idea. He was one of the first scientists to think of Earth as very old. Hutton was a naturalist and a farmer in Scotland. He observed how the landscape on his farm gradually changed over the years. Hutton thought that the processes responsible for changing the landscape on his farm could also shape Earth's surface. For example, he thought that erosion caused by streams on his farm could also wear down mountains or carve deep canyons. Because he realized that this would take a long time, Hutton proposed that Earth is much older than a few thousand years.

Hutton's ideas are the foundation of a principle called uniformitarianism. The principle of **uniformitarianism** states that geologic processes that occur today are similar to those that have occurred in the past. In other words, the same processes that we see today have been occurring since Earth formed.

THREE-DIMENSIONAL THINKING

Scientists use the principle of uniformitarianism to **interpret** Earth's history. Suppose you discover a rock from an ancient beach. Now imagine you are standing on that ancient beach. What do you think you would you see? **Explain** how your answer relates to the principle of uniformitarianism.

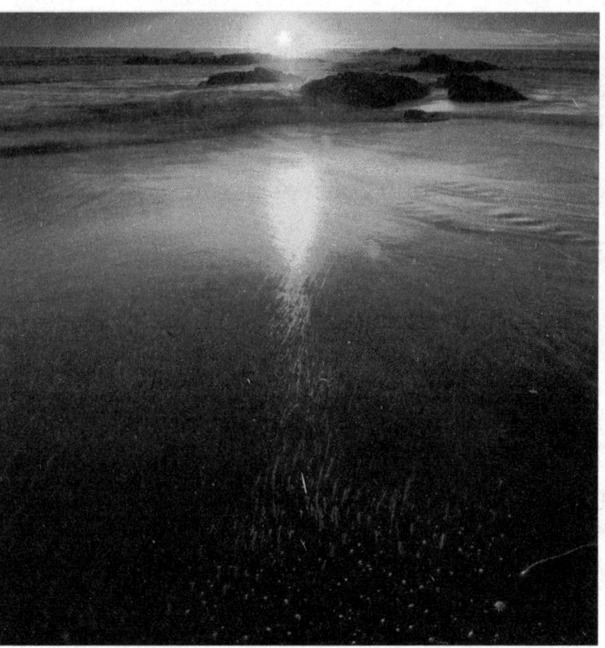

Because of uniformitarianism, scientists can learn about Earth's past by studying the present. One way to do this is by studying the order in which geologic events occurred using a method called relative-age dating. This does not allow scientists to determine the **absolute age,** or the actual age of the events. But it gives scientists a clearer understanding about geologic events in Earth's history.

What is relative-age dating?

Geologists—the scientists who study Earth and rocks—use rock layers and the fossils within to build a record of Earth's geologic history. How do rock layers form? Remember that weathering breaks rocks exposed at Earth's surface into smaller pieces called sediment. Over time, gravity, water, and wind carry sediment downhill and deposit it in low areas. Eventually, layers of sediment form. The increasing weight of the sediment slowly causes the layers to compress, forming layers of rock called strata.
Can you model rock layers?

What do you think happened to these rock layers?

Safety 

Materials

disposable polystyrene meat tray (3)

Procedure

1. Read and complete a lab safety form.
2. Work in groups of 3–4.
3. Break a meat tray in half. Place the two pieces on a flat surface so that the broken edges touch one another.
4. Break another meat tray in half in the same way so that the break is consistent between the two meat trays. Place the two pieces directly on top of the first broken meat tray.
5. Place a third, unbroken meat tray on top of the two broken meat trays. Sketch your model below.

6. Follow your teacher's instructions for proper cleanup.

Analyze and Conclude

7. How do you think your model resembles a rock formation?

8. If you observed rock layers that looked like your model, what would you think might have caused the break only in the two bottom layers? Why might the top layer not be broken?

9. Which layer in your model do you think is the youngest? Which do you think is the oldest? Explain your reasoning.

Relative Age In the *Layers Rock!* lab you interpreted the ages of the layers in comparison to each other. This is called relative age. **Relative age** is the age of rocks and geologic features compared with other rocks and features nearby.

Geologists have developed a set of principles to organize rock layers according to their relative ages. Using the principles of relative-age dating, geologists can determine if rocks or features are younger or older than the rocks or features around them. Let's investigate.

INVESTIGATION

Relatively Speaking...

Analyze the image below.

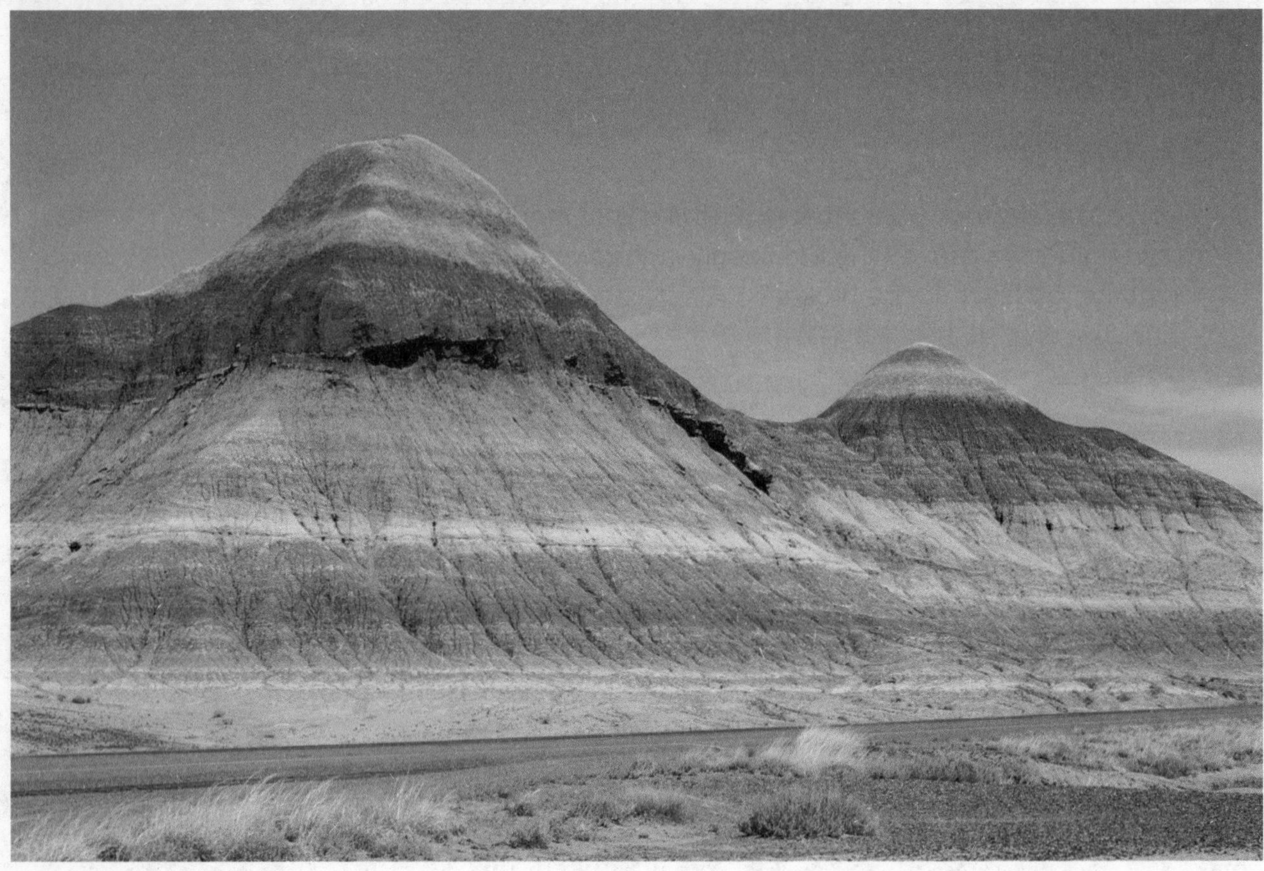

1. Do you think all of the rock layers in the picture formed at the same time? Why or why not?

2. If you think the rocks formed at different times, which layers are the oldest and which are the youngest? Explain.

Now, study this image.

3. Do you think the small pieces of rock in the picture formed at the same time as the rock containing the pieces? Why or why not?

4. If you think the rock formed at different times, which is oldest—the rock containing the pieces, or the pieces of rock? Explain.

Investigation, continued

Now, try to interpret this image.

5. Notice the large fault cutting across the rock layers. Do you think the fault and the rock layers are the same age? Why or why not?

6. If you think the fault and the rock formed at different times, which is oldest—the fault or the rocks? Explain.

GO ONLINE Finally, watch the animation *Relative-Age Dating*. Then answer the questions that follow.

7. What are the principles of relative-age dating?

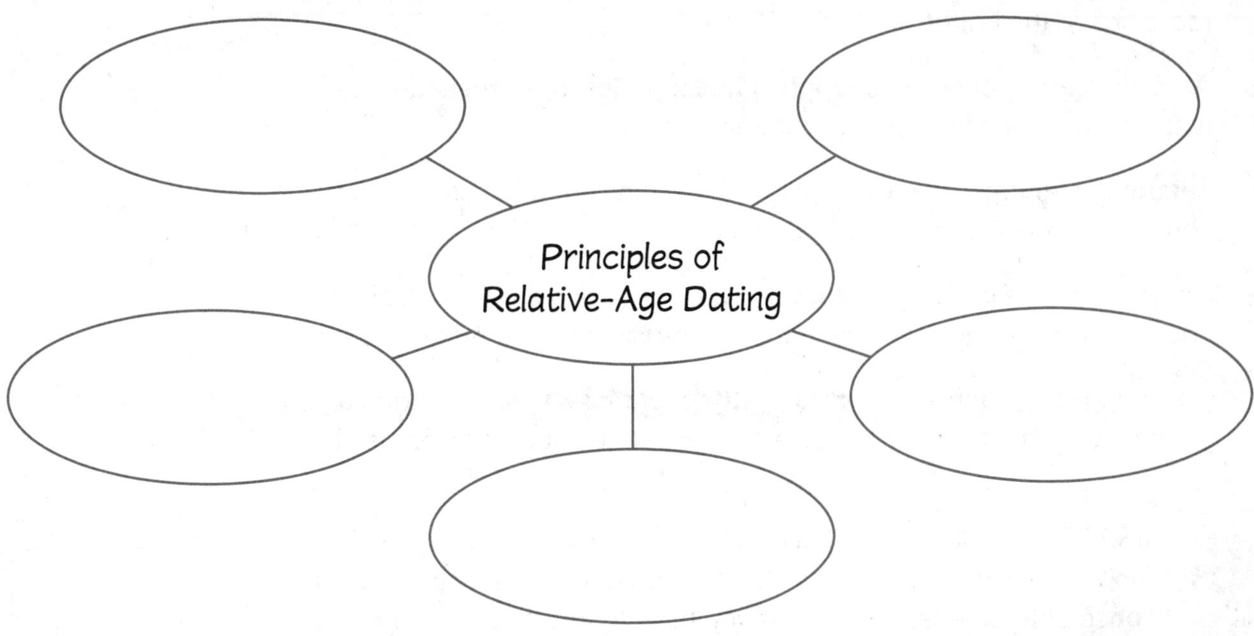

8. Which principle did you apply in Step 2 of the investigation?

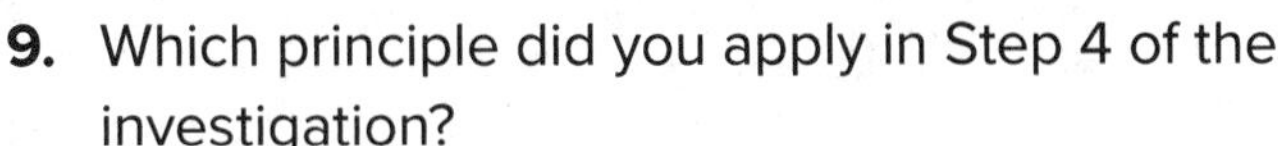

Using this principle, assign the rock layers in the diagram to the right their relative ages from oldest (*1*) to youngest (*4*).

9. Which principle did you apply in Step 4 of the investigation?

Using this principle, label the older feature *1*, and the younger feature *2* on the diagram to the right.

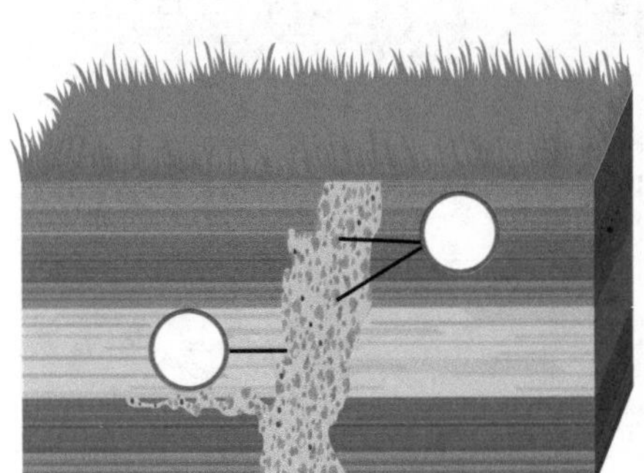

10. Which principle did you use in Step 6 of the investigation?

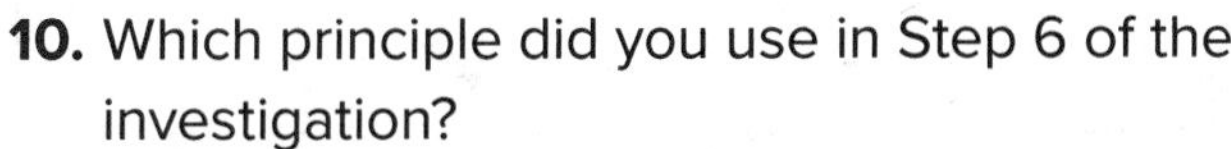

Using this principle, label the rock layers and features from oldest (*1*) to youngest (*6*).

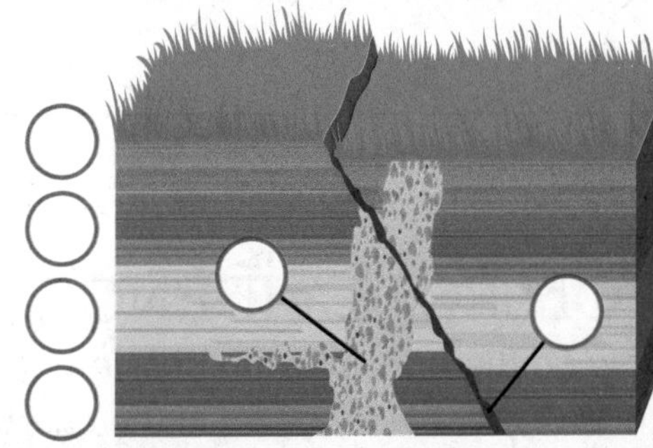

The Principles of Relative-Age Dating As you just discovered in the *Relatively Speaking...* investigation, there are five principles that geologists use to assign relative ages to rocks and features.

- **Superposition** is the principle that in undisturbed rock layers, the oldest rocks are on the bottom.
- According to the principle of **original horizontality,** most rock-forming materials are deposited in horizontal layers.
- **Lateral continuity** is the principle that sediments are deposited in large, continuous sheets in all lateral directions.
- According to the principle of **inclusions,** if one rock contains pieces of another rock, the rock containing the pieces is younger than the pieces.
- According to the principle of **cross-cutting relationships,** if one geologic feature cuts across another feature, the feature that it cuts across is older.

THREE-DIMENSIONAL THINKING

Models are often used to study and represent **large-scale** time and space phenomena. Using what you have learned about the principles of relative-age dating, create a sketch modeling each principle below.

COLLECT EVIDENCE

How do the principles of relative-age dating help scientists "read" rocks? Record your evidence (A) in the chart at the beginning of the lesson.

What can the fossil record tell us about Earth's history?

You just discovered how the principles of relative-age dating allow geologists to determine if rocks or features are younger or older than the rocks or features around them. But did you know that the rocks themselves can hold clues about their ages?

LIFE SCIENCE Connection Some of the most obvious clues found in rocks are **fossils**—the remains or traces of ancient living things. How can geologists use fossils to infer the ages of rocks and events in Earth's history? Let's dig in.

INVESTIGATION

Clues to Earth's Past

Analyze the image below. Then answer the questions that follow.

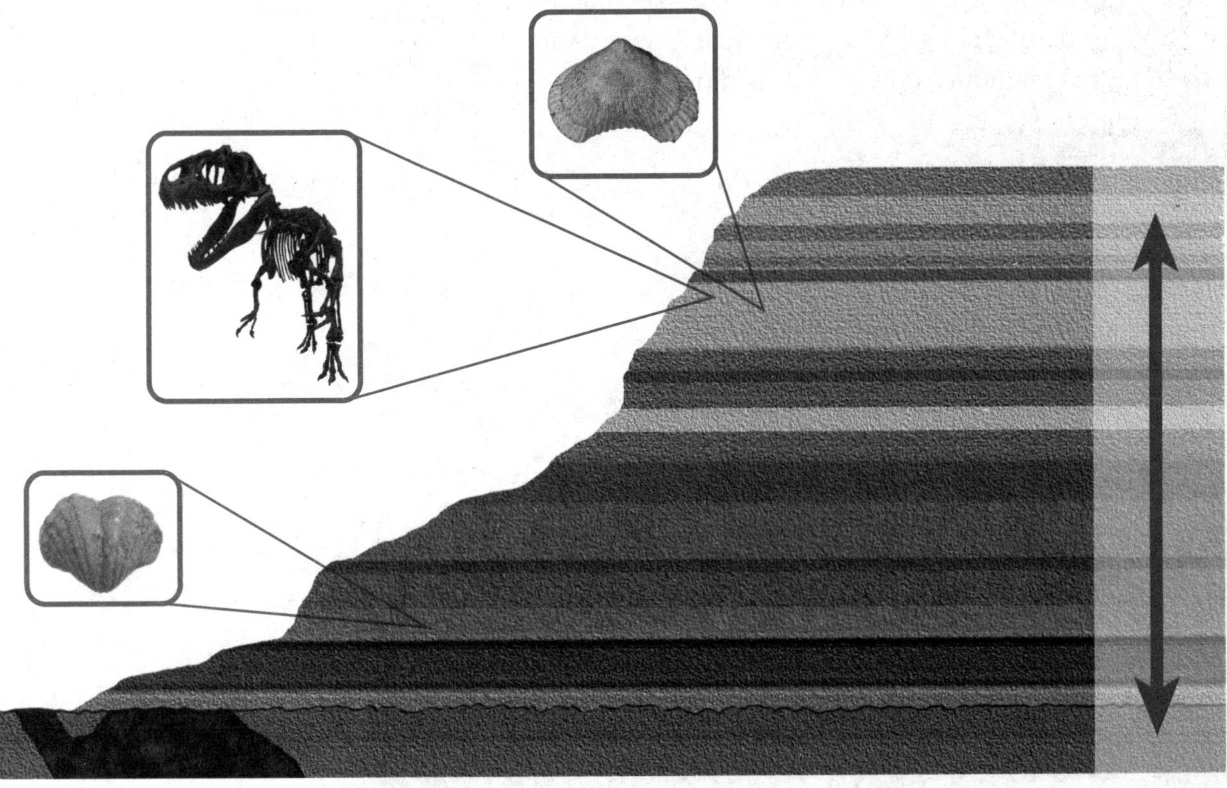

Note: Fossils are not to scale.

1. Using the principle of superposition, determine which rock layers are the oldest, and which rock layers are the youngest. Label the space above and below the arrow heads *Oldest* and *Youngest* accordingly.

Investigation, continued

2. Imagine you discovered the fossils in the rock layers on the previous page. What do you notice about the complexity of the fossils in the rocks? Where are the more simple fossils located? Where is the more complex fossil located?

____

____

____

____

3. Make a claim about how the complexity of fossils in the fossil record can provide clues to the relative ages of rock units.

____

____

____

Now, study the image below. Notice the fossil trilobites in the rock layers at the bottom of the sequence. As you move upward in the sequence you'll see fossils of mammals and terrestrial dinosaurs.

Note: Fossils are not to scale.

4. When do trilobites disappear from the rock sequence? Label this point *Disappearance of trilobites* in the white space to the right of the image.

5. When do mammals and dinosaurs appear in the rock sequence? Label this point *Appearance of mammals and dinosaurs.*

6. What can you conclude about the layers that contain the trilobites? Are they older or younger than the layers that contain the mammals and dinosaurs?

7. When do the dinosaurs disappear from the rock sequence? Label this point *Disappearance of dinosaurs* in the white space to the right of the image.

8. What can you conclude about the layers that contain only the mammals? Are they older or younger than the layers that contain the dinosaurs?

9. Make a claim about how the appearance and disappearance of organisms in the fossil record provides relative dates.

GO ONLINE for additional opportunities to explore!

LIFE SCIENCE Connection Want to learn more about fossils? Then perform one of the following activities.

☐ **Investigate** clues about how organisms lived in the **Lab** *What can trace fossils show?*

OR

☐ **Explore** ancient organisms by studying fossil remains in the **Lab** *How is a fossil a clue?*

The Fossil Record While studying the fossils in rock layers, as you just did in the *Clues to Earth's Past* investigation, early geologists recognized three things. First, older rocks contained only fossils of small, relatively simple life-forms. Younger rocks contained these fossils as well as fossils of other, more complex organisms. Second, sometimes fossils in one rock layer did not appear in the rock layers right above it. It seemed as though the organisms that lived during that period of time had disappeared suddenly. Sudden changes in the fossil record represent **mass extinctions**—times when many species on Earth died or became extinct within a short period of time.

And finally, fossils and the rocks they are within can be used to determine what the environment of an area was like long ago. Let's look at an example.

INVESTIGATION

Reconstructing Past Environments

GO ONLINE Watch the video *Secrets of the Origin of Life*.

1. What does the earliest evidence of life look like in the fossil record?

____

____

____

2. What do these rocks tell scientists about the environment in which they formed?

____

____

____

Ancient Environments Scientists use the principle of uniformitarianism to learn about ancient organisms and the environments where ancient organisms lived. For example, a fossil of an ancient coral is evidence that the surrounding location was a marine environment when the coral was alive.

COLLECT EVIDENCE

How does the fossil record help scientists piece together Earth's history? Record your evidence (B) in the chart at the beginning of the lesson.

A Closer Look: Drilling Into the Past

Did you know fossils from the seafloor can provide evidence of past climates? Scientists use core samples of the seafloor to gather data about Earth's history. Each layer holds grains of dust, minerals, and often times pollen, that can provide information about the environment in which they were deposited. Many layers also contain fossils of the organisms that lived in the ocean during that time.

The most common types of fossils analyzed in deep-sea sediment cores are foraminifera. Foraminifera are tiny, unicellular organisms less than the size of a grain of sand. When they grow, foraminifera incorporate oxygen from the surrounding water into their shells.

Oxygen comes in two important varieties for paleoclimate research: heavy (^{18}O) and light (^{16}O). Water vapor molecules containing the heavy isotope condense more readily than water vapor molecules containing the light isotope. When climates are cooler, the water vapor containing ^{18}O rains out of the atmosphere. The wind carries water vapor containing the lighter ^{16}O toward Earth's poles, where it eventually condenses and falls onto the ice sheets where it remains trapped. As a result, the oceans develop increasingly higher concentrations of heavy oxygen during cooler time periods. In general, foraminifera shells contain more heavy oxygen during glacial periods.

It's Your Turn

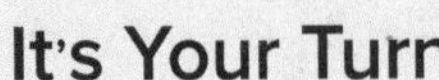

Research the latest expeditions of the *JOIDES Resolution,* a scientific drilling ship sponsored by the Joint Oceanographic Institutions for Deep Earth Sampling. Develop a poster showing where the ship has traveled as well as the purpose of its expeditions.

Foraminifera
LM Magnification: unavailable

LESSON 1

Review

Summarize It!

1. **Organize** In the space below, create a graphic organizer summarizing how the relative ages of rocks, geologic features, and events are determined using the rock and fossil records.

Three-Dimensional Thinking

Imagine you are a geologist. You have been asked to analyze and interpret the rock sequence below. Your task is to determine the relative ages of the rocks.

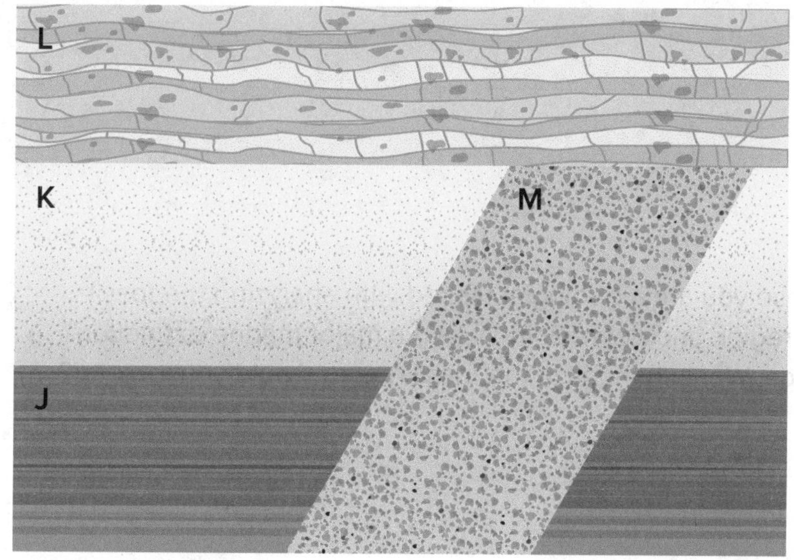

2. Order the features in the illustration from oldest to youngest.

A JKLM

B MJKL

C JKML

D MLKJ

3. Which geologic principle must be assumed to determine the relative age of M?

A cross-cutting relationships

B superposition

C original horizontality

D inclusions

Real-World Connection

4. **Contrast** What is your absolute age? What is your relative age compared to two of your classmates? How do these ages contrast?

5. **Predict** Suppose you are a geologist tasked with analyzing the fossil record in your region and determining when a major marine extinction event took place. What changes in the fossil record would you look for?

Still have questions?

Go online to check your understanding about the rock and fossil records.

REVISIT

Do you still agree with the answer you chose at the beginning of the lesson? Return to the Science Probe at the beginning of the lesson. Explain why you agree or disagree with that answer now.

EXPLAIN THE PHENOMENON

Revisit your claim about how the sequence of past geologic events can be determined. Review the evidence you collected. Explain how your evidence supports your claim.

START PLANNING

STEM Module Project Science Challenge

Now that you've learned about how geologists analyze the rock and fossil records, go to your Module Project to start planning your video blog. Keep in mind that you want to explain how evidence from rock strata is used to organize Earth's vast geologic past.

PAGE KEELEY SCIENCE PROBES

LESSON 2 LAUNCH

Scientific Explanations

An explanation helps provide answers to a question a scientist might be wondering about. Which of the following do you think involves providing a scientific explanation?

A hypothesis

B scientific theory

C scientific law

D hypothesis and scientific theory

E Scientific theory and scientific law

F hypothesis, scientific theory, and scientific law

G None of the above. An explanation is something else.

Explain your thinking. Describe how explanations are used in science.

____

____

____

____

You will revisit your response to the Science Probe at the end of the lesson.

LESSON 2

Building a Time Line

ENCOUNTER THE PHENOMENON

How do geologists know that rock layers that are far apart from each other formed at the same time?

GO ONLINE

Watch the animation *The Grand Staircase* to see this phenomenon in action.

Did you know that the layer that rims the top of the Grand Canyon in Arizona is also found more than 100 km away at the bottom of Zion National Park in Utah? With your partner, decide how geologists might know that these layers, which are far apart from each other, formed at the same time. How would this information help in building a time line of Earth's history? Record your ideas below.

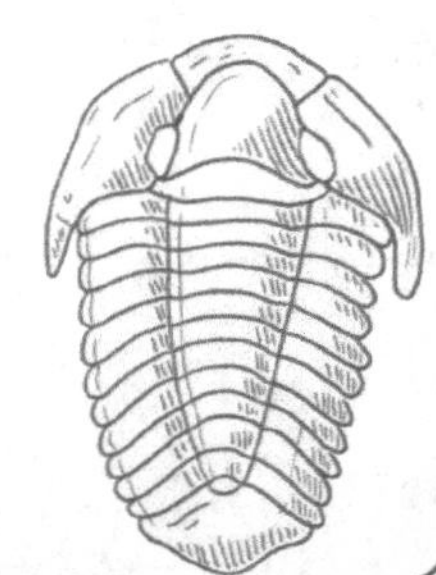

EXPLAIN THE PHENOMENON

Are you starting to get some ideas about the clues scientists use to connect rock layers that are far apart, and how this might help build a record of Earth's past? Use your observations about the phenomenon to make a claim about how geologists build a record of Earth's geologic history.

CLAIM

Geologists build a record of Earth's geologic history by...

COLLECT EVIDENCE as you work through the lesson.

Then return to these pages to record your evidence.

EVIDENCE

A. What evidence have you discovered to explain how geologists fill gaps in the rock record?

B. What evidence have you discovered to explain how major events help geologists organize Earth's history?

MORE EVIDENCE

C. What evidence have you discovered to explain the geologic time scale?

When you are finished with the lesson, review your evidence. If necessary, based on the evidence, revise your claim.

REVISED CLAIM

Geologists build a record of Earth's geologist history by...

Finally, explain your reasoning for how and why your evidence supports your claim.

REASONING

The evidence I collected supports my claim because...

What happens when the rock and fossil records are not complete?

You learned in Lesson 1 that geologists use rock layers and the fossils within to build a record of Earth's geologic history. A complete rock record in one place, however, does not exist. Weathering, erosion, volcanism, and other processes are constantly changing Earth's surface. This makes it difficult to find a sequence of rock layers that haven't been disturbed. Sometimes, the record of a past event or time is completely eroded away! Let's dig in.

Want more information?
Go online to read more about the development of the geologic time scale.

Go to the Foldables® library to make a Foldable® that will help you take notes while reading this lesson.

INVESTIGATION

Gaps in the Rock Record

1. Analyze the three photos below. Notice how the arrows point to a line between different rocks. How would you describe the rock below the arrows versus the rock above? Write your descriptions next to each image.

A

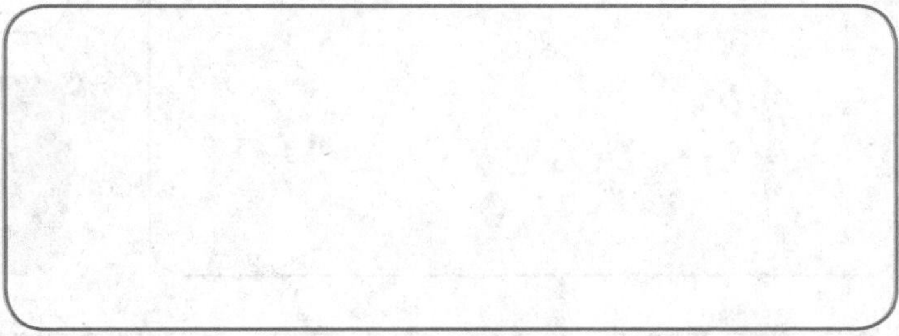

B

C

2. The arrows in the images on the previous page point to the surfaces where rock has eroded away, producing breaks, or gaps, in the rock record, called **unconformities** (un kun FOR muh tees). There are three types of unconformities, described below. Can you identify which photo from the previous page each type of unconformity describes? Label the figures *A, B,* or *C* to correspond with the matching photos.

Younger sedimentary rock

Older sedimentary rock

◀ When horizontal layers of sedimentary rock are deformed during mountain building or other geologic events, they are usually uplifted and tilted. During this process, the layers are exposed to weathering and erosion. If horizontal layers of sedimentary rock are later laid down on top of the tilted, eroded layers, the resulting unconformity is called an angular unconformity.

Younger sedimentary rock

Older sedimentary rock

◀ When a horizontal layer of sedimentary rock overlies another horizontal layer of sedimentary rock that has been eroded, the eroded surface is called a disconformity. Disconformities can be easy to identify when the eroded surface is uneven. When the eroded surface is smooth, disconformities are often hard to see.

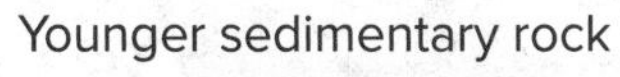

Older igneous rock

◀ When a layer of sedimentary rock overlies a layer of igneous or metamorphic rock, such as granite or marble, the eroded surface is easier to identify. This kind of eroded surface is called a nonconformity. A nonconformity indicates a gap in the rock record during which rock layers were uplifted, eroded at Earth's surface, and new layers of sedimentary rock formed on top.

Unconformities As you just discovered, when new sediment is deposited on top of old, eroded rock layers, the eroded surface represents a gap in the rock record. Unconformities could represent a few hundred years, a million years, or even billions of years. So how can geologists organize Earth's history using the rock and fossil records when parts are buried or missing? Let's investigate.

How can we fill gaps in the rock record?

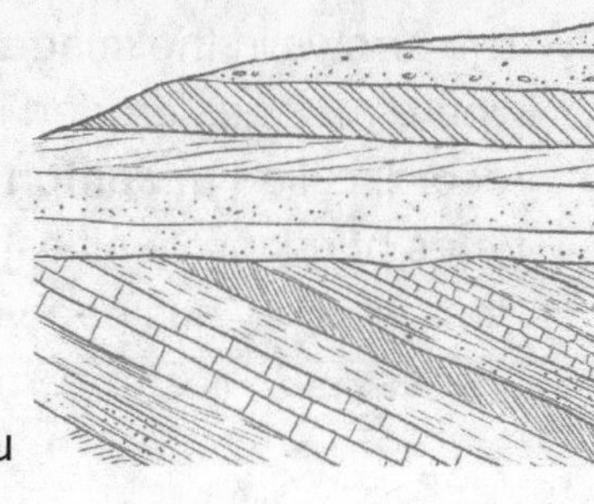

Geologists fill in gaps in the rock record by matching of rock layers or fossils from separate locations. This is particularly helpful when unconformities occur, or when rocks are buried underground. Can you match layers of rock?

INVESTIGATION

It's a Match!

Imagine you are a geologist and you have been asked to match the rock sequences below that have been separated by a canyon.

1. Which rock layers have matching rock types in the two rock sequences shown here? Draw lines connecting the layers that match. The first has been done for you.

2. Which principle of relative-age dating did you use when correlating the rock layers? Explain.

3. The oldest rock layer has been labeled 1. Assign relative ages to the remaining rock layers. Which geologic principle did you use to determine this? Explain.

Now analyze these two rock formations from different continents. Answer the questions that follow.

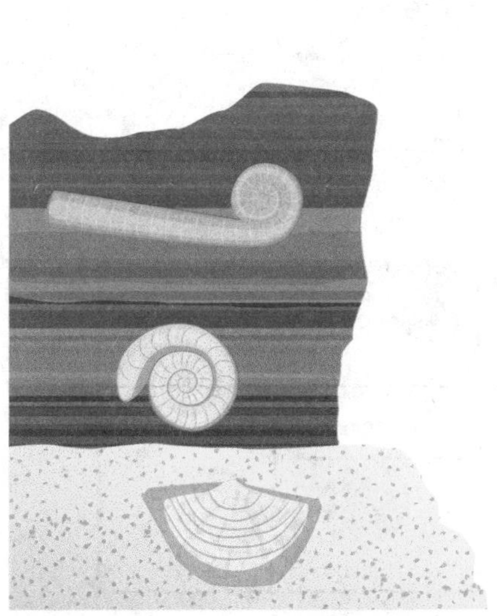

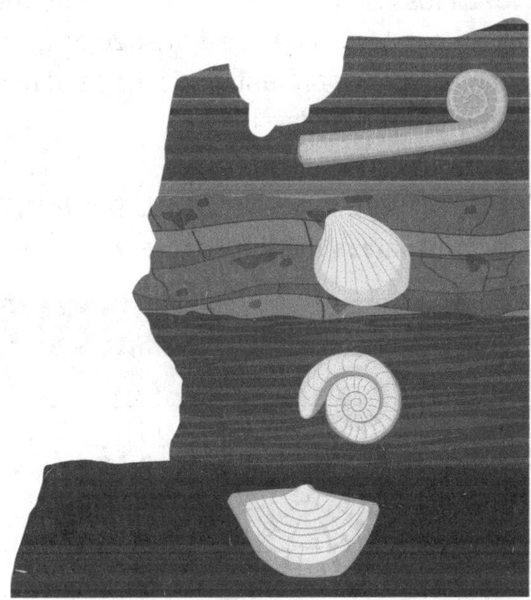

4. Do any of the rock layers match based on their rock type? Why or why not?

5. Which fossils match in the two rock formations shown here? Draw lines connecting the layers that have matching fossils.

6. Notice that rock layers are missing from some of the rock sequences. What must have occurred in these locations?

Correlation As you just explored in the *It's a Match!* investigation, geologists fill gaps in the rock record through **correlation** (kor uh LAY shun)—the matching of rock layers or fossils exposed in one geographic region to similar layers or fossils exposed in other geographic regions. If the rock formations are very far apart or even on different continents, geologists often rely on fossils. If two or more rock formations contain fossils of about the same age, scientists can infer that the formations are also about the same age, even if the rock types are different.

Not all fossils are useful in determining the relative ages of rock layers. The most useful fossils represent species that existed on Earth for a short length of time, were abundant, and inhabited many locations. These fossils are called **index fossils.**

THREE-DIMENSIONAL THINKING

Analyze the time **scales** for the following fossils.

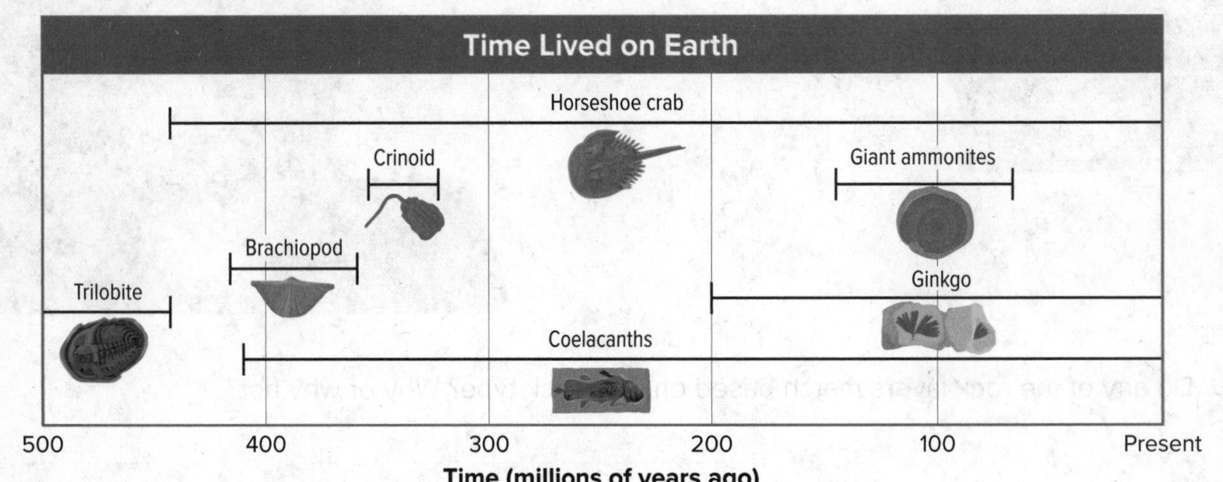

Which fossils appear to be index fossils? **Explain** your choices.

____

____

The correlation of fossils and rock layers aids in the relative dating of rock sequences and helps geologists understand the history of larger geographic regions. It is largely through correlation that geologists have constructed Earth's time line.

COLLECT EVIDENCE

How do geologists fill gaps in the rock record? Record your evidence (A) in the chart at the beginning of the lesson.

STEM CAREER Connection

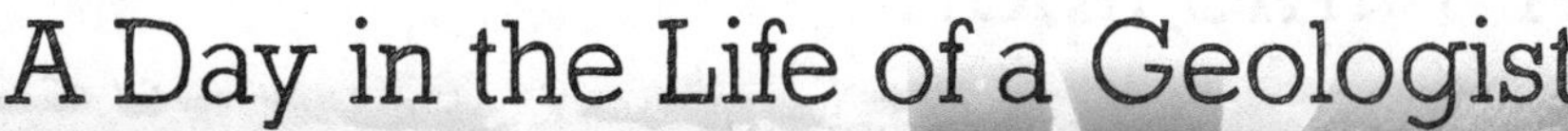

A Day in the Life of a Geologist

Geologists produce descriptions, maps, and other data needed to reconstruct Earth's history and to determine how this history is important to us today. The traditional view of a geologist is of a field geologist, a person outdoors with a backpack, hiking boots, and a rock hammer, traversing a scenic mountain range. They also investigate questions using laboratories, outdoor experimental facilities, numerical modeling, and other strategies.

During field studies, geologists observe various aspects of the natural environment, record these observations, and propose explanations for these observations. A common goal is to understand the area's geologic processes and history, commonly by producing a geologic map and geologic cross sections.

It's Your Turn

WRITING Connection Walter Alvarez was a young paleontologist, a geologist specializing in using fossils to decipher the history of Earth. Research the Alvarez hypothesis about the extinction of the dinosaurs 66 million years ago. What clues did he find in the rock record? Write a short blog on what you find.

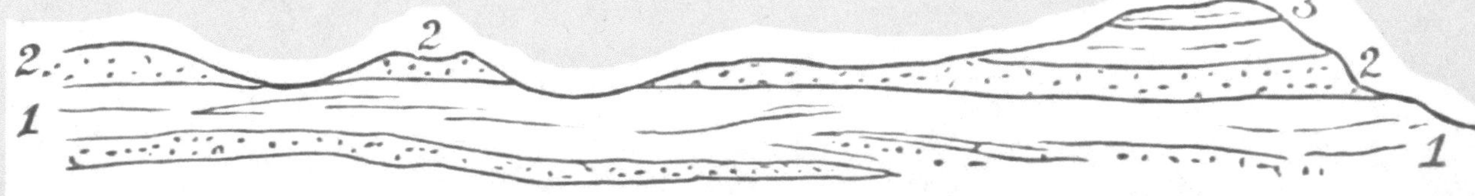

How do major geologic events help build Earth's time line?

Sometimes major events in Earth's history can leave a distinctive layer in the rock record. Let's take a look.

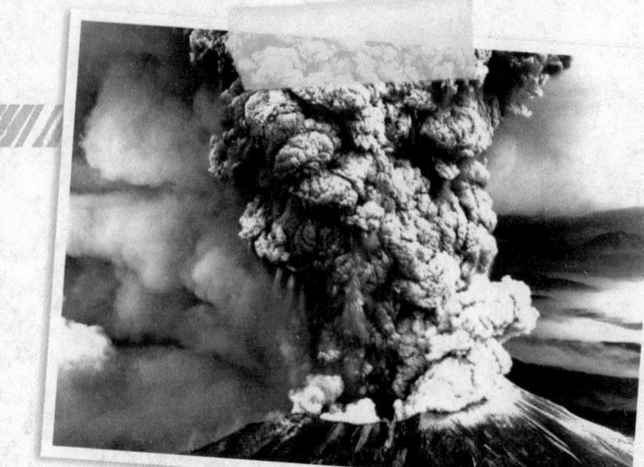

INVESTIGATION

Ashes, Ashes, We All Fall Down

Examine the photograph of the 1980 eruption of Mount St. Helens.

1. What do you think happens to the ash from huge volcanic eruptions like Mount St. Helens?

2. Assuming the ash layers in the figure below have been dated as shown, what conclusions can you draw about the ages of each of the layers above and below the ash layers?

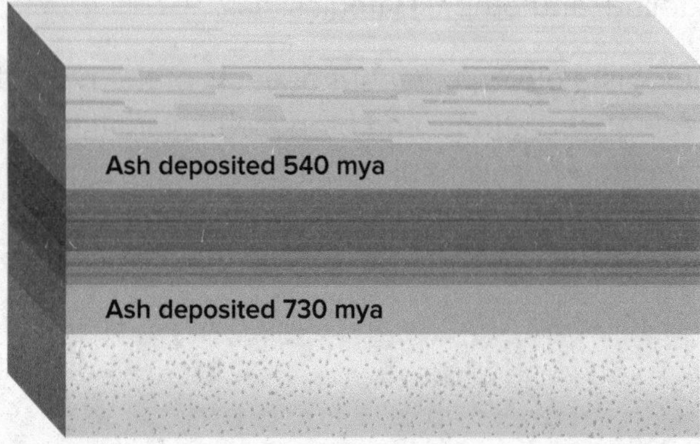

THREE-DIMENSIONAL THINKING

Scientists hypothesize that a meteorite impact might have caused the mass extinction that occurred when dinosaurs became extinct. Evidence for this impact is in a clay layer containing the element iridium in rocks around the world. Iridium is rare in Earth rocks but common in meteorites.

Analyze the image below.

Nearly all fossils below the iridium layer in Earth's rocks are different from those above, indicating that a mass extinction occurred.

1. Would you expect find dinosaur fossils in undisturbed rocks that are above the iridium layer? **Explain** your answer.

2. How does the **structure** of a distinctive layer that is used as a marker determine its **function?**

Key Beds As you just examined, a large meteorite strike, volcanic eruption, or other major event can leave a unique layer in the rock record. Because these types of layers are easy to recognize, they help geologists correlate rock formations in different geographic areas where layers are exposed. A rock or sediment layer used as a marker in this way is called a **key bed.** Using the principle of superposition, geologists know that the layers above a key bed are younger than the layers below it.

Not all of Earth's major events leave distinctive key beds, but they do leave evidence in the rock record. In the following investigation, you will explore how the type and order of rock layers can be used as evidence for the formation of mountains.

The Riddle of the Rocks

GO ONLINE Watch the video *Layers of Rock*.

1. **HISTORY Connection** How did Roderick Murchison's and Charles Lapworth's views of the formation of Scotland's mountains differ?

2. How can rock layers tell the story of mountain building?

Major Geologic Events Using a combination of the principles of relative-age dating, the fossil record, and evidence of major events in Earth's history, geologists can determine the relative order of events. In doing so, a time line of Earth's long past can be constructed.

COLLECT EVIDENCE

How can relative ages be determined using major events? Record your evidence (B) in the chart at the beginning of the lesson.

Read a Scientific Text

Earth has a long history. Rocks around the world contain evidence that vast mountains were uplifted and eroded away and that seas advanced and retreated across the land many times.

CLOSE READING

Inspect

Read the passage *Relative Time Scale.*

Find Evidence

Reread the second paragraph. Circle the evidence for the formation of mountain chains and ocean basins.

Make Connections

Communicate With your partner, discuss the similarities between how humans organize history and how geologic time is organized. Can you provide an example of a division of time in your life?

Relative Time Scale

William "Strata" Smith, a civil engineer and surveyor, was well acquainted with areas in southern England where "limestone and shales are layered like slices of bread and butter." His hobby of collecting and cataloging fossil shells from these rocks led to the discovery that certain layers contained fossils unlike those in other layers. Using these key or index fossils as markers, Smith could identify a particular layer of rock wherever it was exposed. Because fossils actually record the slow but progressive development of life, scientists use them to identify rocks of the same age throughout the world.

From the results of studies on the origins of the various kinds of rocks (petrology), coupled with studies of rock layering (stratigraphy) and the evolution of life (paleontology), geologists reconstruct the sequence of events that has shaped the Earth's surface. Their studies show, for example, that during a particular episode the land surface was raised in one part of the world to form high plateaus and mountain ranges. After the uplift of the land, the forces of erosion attacked the highlands and the eroded rock debris was transported and redeposited in the lowlands. During the same interval of time in another part of the world, the land surface subsided and was covered by the seas. With the sinking of the land surface, sediments were deposited on the ocean floor. The evidence for the pre-existence of ancient mountain ranges lies in the nature of the eroded rock debris, and the evidence of the seas' former presence is, in part, the fossil forms of marine life that accumulated with the bottom sediments.

Such recurring events as mountain building and sea encroachment, of which the rocks themselves are records, comprise units of geologic time even though the actual dates of the events are unknown. By comparison, the history of mankind is similarly organized into relative units of time. We speak of human events as occurring either B.C. or A.D.—broad divisions of time. Shorter spans are measured by the dynasties of ancient Egypt or by the reigns of kings and queens in Europe. Geologists have done the same thing to geologic time by dividing the Earth's history into Eras—broad spans based on the general character of life that existed during these times—and Periods— shorter spans based partly on evidence of major disturbances of the Earth's crust.

Source: United States Geological Survey

What is the geologic time scale?

To organize events in your life, you use different units of time, such as days, weeks, months, and years. Geologists organize Earth's past in a similar way. They developed a model of Earth's history from its origin 4.6 billion years ago to the present called the **geologic time scale,** shown on the opposite page. Time units on the geologic time scale are thousands and millions of years long—much longer than the units you use to organize events in your life.

INVESTIGATION

It's About Time!

1. Study the geologic time scale on the opposite page. Where is the oldest division of time located on the time scale? Where is the youngest division of time?

2. Why do you think the scale is laid out in this manner?

3. Do the boundaries between the units appear at regular intervals? Why do you think the units are unequal in length?

The Geologic Time Scale Geologists developed the geologic time scale to help them correlate rock units across countries and continents and to have a standard model and vocabulary for describing geologic time.

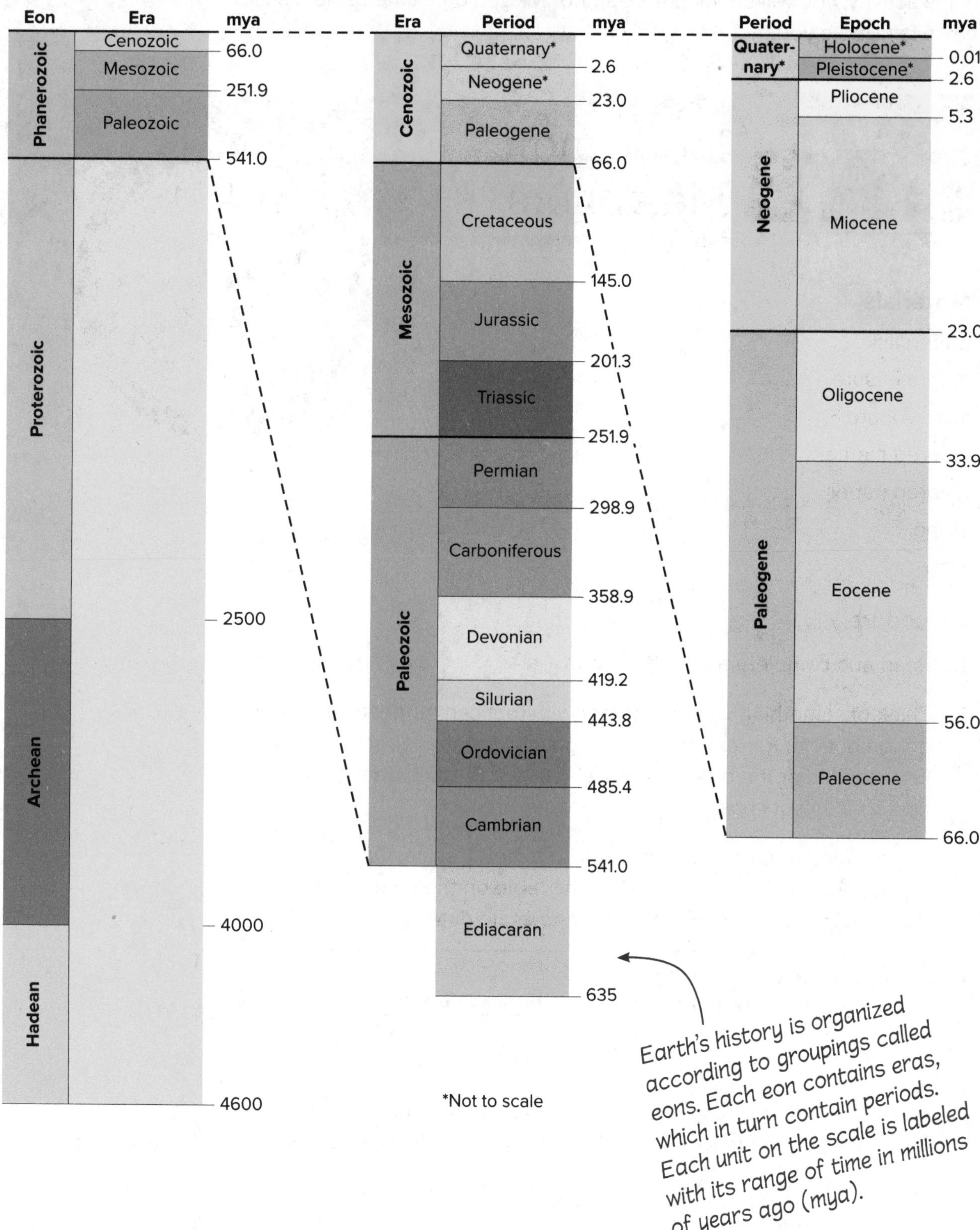

COLLECT EVIDENCE

How does the geologic time scale organize Earth's history? Record your evidence (C) in the chart at the beginning of the lesson.

Relating Time Scales Earth formed approximately 4.6 billion years ago. But how long is 4,600,000,000 years? It is difficult to comprehend time that extends so far into the past unless you can relate it to your own experience. In this activity, you will develop a metaphor for geologic time using a scale that is familiar to you.

Materials

meterstick
tape measure
posterboard
colored markers
colored paper
string
maps

Procedure

1. Read and complete a lab safety form.
2. Think of something you are familiar with that can model a long period of time. For example, you might choose the length of a football field or the distance between two U.S. cities on a map—one on the east coast and one on the west coast.
3. Make a model of your metaphor using a metric scale. On your model, display the events listed in the table on the next page. Use the equation below to generate true-to-scale dates in your model.

$$\frac{\text{Known age of past event (years before present)}}{\text{Known age of Earth (years before present)}} = \frac{X \text{ time scale unit location}}{\text{Maximum distance or extent of metaphor}}$$

Example: To find where "first fish" would be placed on your model if you used a meterstick (100 cm), set up your equation as follows:

$$\frac{500{,}000{,}000 \text{ years}}{4{,}600{,}000{,}000 \text{ years}} = \frac{X \text{ (location on meterstick)}}{100 \text{ cm}}$$

4. In your Science Notebook, keep a record of all the math equations you used. You can use a calculator, but show all equations.

Analyze and Conclude

5. **MATH Connection** Calculate the percentage of geologic time modern humans have occupied.

Some Important Approximate Dates in the History of Earth	
MYA	**Event**
4,600	Origin of Earth
3,700	Oldest evidence of life
500	First fish
320	First reptiles
252	Permian extinction event
220	Mammals and dinosaurs appear
145	Atlantic Ocean forms
66	Cretaceous extinction event
6	Human ancestors appear
2.6	Pleistocene Ice Age begins
0.1	*Homo sapiens* appear
0.00052	Columbus lands in New World
??	Your birth date

6. What other milestone events in Earth's history, other than those listed on the table, could you include on your model?

7. The Earth events on your model are based mostly on fossil evidence. How are fossils useful in understanding Earth's history? How are they useful in the development of the geologic time scale?

Summarize It!

1. **Outline** your understanding of the development of the geologic time scale using the chart below. The topic is *Building a Time Line*. Choose three central ideas you learned about in this lesson and provide specific details about each main idea.

Topic: ______________________________

Main Ideas	*Specific Details*

Three-Dimensional Thinking

Through correlation of many different layers of rocks, geologists have determined that Zion National Park, Bryce Canyon, and the Grand Canyon are all part of one layered sequence called the Grand Staircase, shown below. Some of these layers are buried underground.

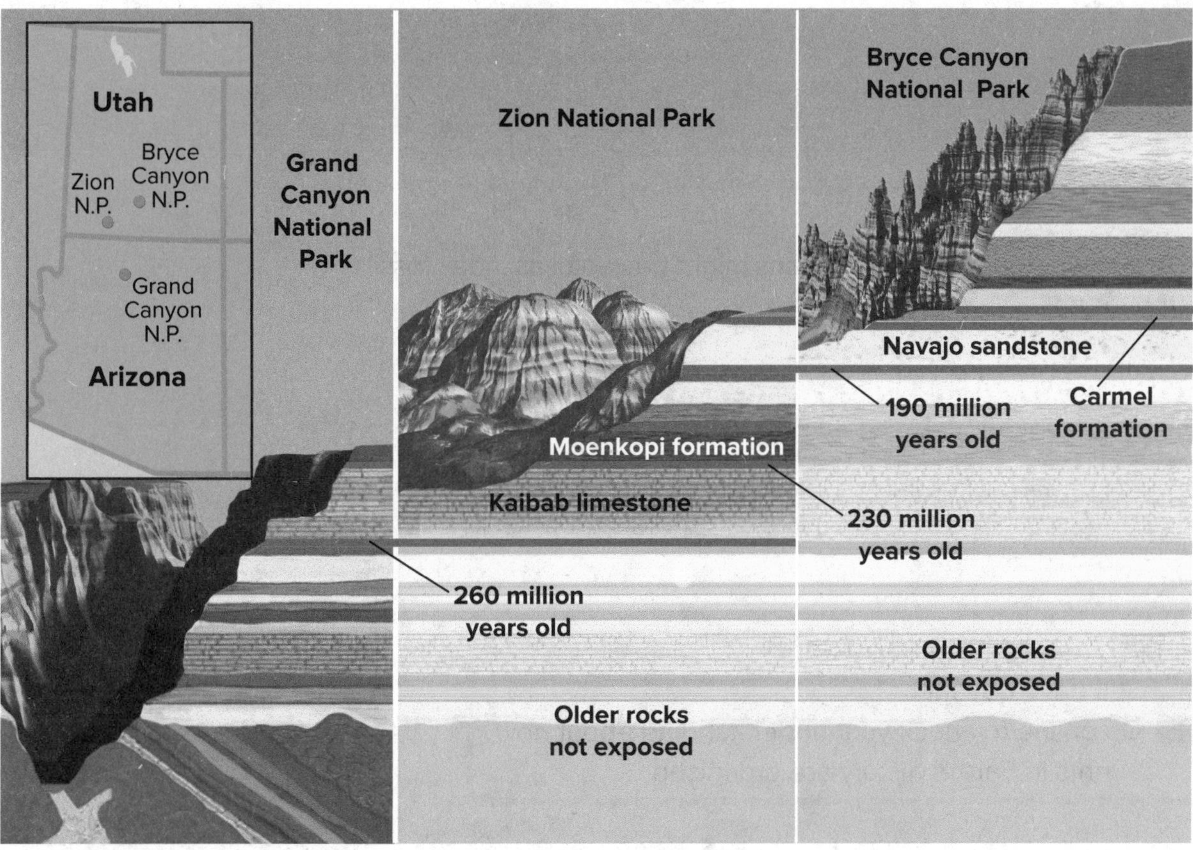

2. Infer the makeup of the buried layer below Zion's Kaibab layer.

   A The layer below the Kaibab layer at Zion National Park would likely be the same layer that is immediately above the Kaibab layer at Bryce Canyon National Park.

   B The layer below the Kaibab layer at Zion National Park would likely be the same layer that is immediately below the Kaibab layer at Grand Canyon National Park.

   C The layer below the Kaibab layer at Zion National Park would likely be the same layer that is immediately above the Navajo layer at Bryce Canyon National Park.

   D The layer below the Kaibab layer at Zion National Park would likely be the same layer that is immediately below the Navajo layer at Grand Canyon National Park.

Real-World Connection

3. Evaluate the difference between human time and geologic time.

4. Argue whether you think humans might be useful as index fossils in the future.

Still have questions?
Go online to check your understanding about how events in Earth's history are organized.

REVISIT

Do you still agree with the answer you chose at the beginning of the lesson? Return to the Science Probe at the beginning of the lesson. Explain why you agree or disagree with that answer now.

EXPLAIN THE PHENOMENON

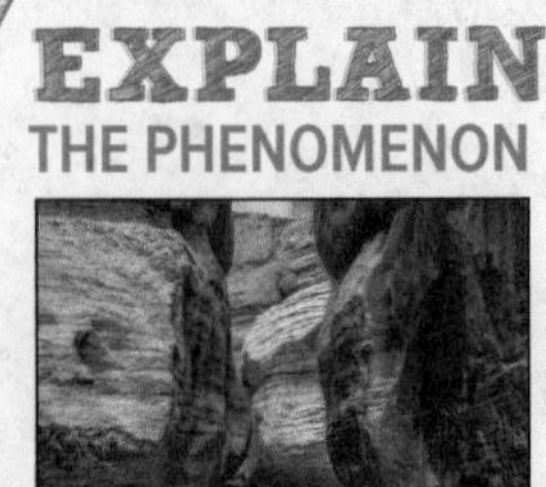

Revisit your claim about how geologists build a record of Earth's geologic history. Review the evidence you collected. Explain how your evidence supports your claim.

PLAN AND FINISH UP

STEM Module Project
Science Challenge

Now that you've learned about the development of the geologic time scale, go back to your Module Project to continue planning your video blog and give your presentation. Your goal is to explain how the geologic time scale is used to organize Earth's history.

STEM Module Project
Science Challenge

History of Rock

Maria is in Ms. Monroe's science class. She and her partner, Sam, are studying a map of the last ice age. Maria looks thoughtful. She raises her hand.

"Yes, Maria?" Ms. Monroe asks.

"How did scientists come up with this map?" Maria asks. "There aren't any written records. What's their evidence?"

"Funny, I was going to ask you the same question," Ms. Monroe says with a smile. "Your homework assignment is to explain how rocks and fossils provide evidence about major events in Earth's history, like the last ice age, and how this evidence is used to organize the geologic time scale. And be creative! Think of a fun way to present your explanation."

Maria and Sam look at one another. That sounds like a lot of work! They'll need your help to create a video blog that explains how evidence from rock strata is used to organize Earth's vast geologic past.

Planning After Lesson 1

As a group, research several major events from Earth's geologic past. Use multiple valid and reliable sources for your research. Decide which event you would like to explore further. Describe your event below. What happened? When did it happen? Where did it happen?

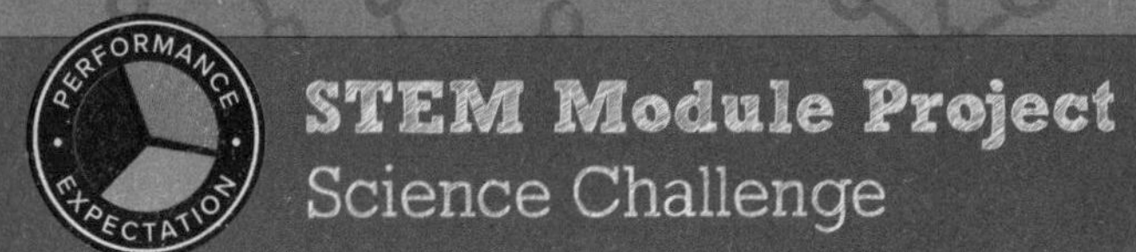

Planning After Lesson 2

Research the evidence used by scientists to establish the relative age of your chosen event. Use your research to complete the table below.

Evidence in the Rock Record	
Evidence in the Fossil Record	

Analyze the data that you've gathered. Use your analysis to make an outline for your video blog. Your blog should explain how evidence from rock strata is used to determine the relative ages of major events in Earth's history, and to organize that history in the geologic time scale. Be sure to identify and describe your evidence. Write your outline for the video blog in your Science Notebook.

Evaluate Your Video Blog

Create your video blog, then use the criteria below to evaluate your work and to describe any revisions you will make.

Does your video blog ...	Yes	No	Describe Revisions to Meet Criteria
...include a detailed description of a major event in the geologic time scale?			
...include an analysis of evidence (correlations, fossils, key beds) from rock strata about the event?			
...relate your analysis to the relative dating of major events in Earth's history and the organization of the geologic time scale?			

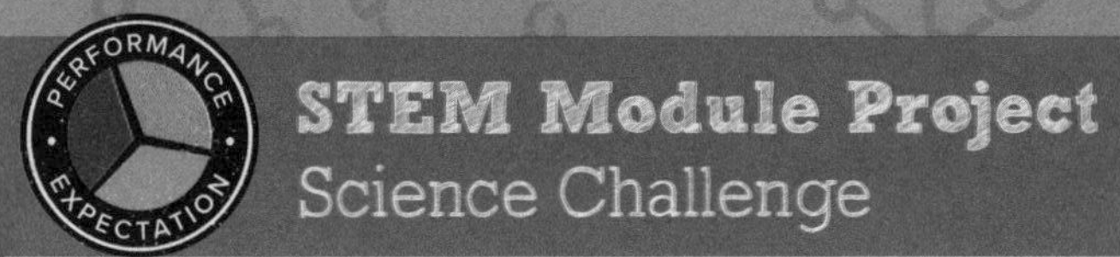

Present Your Video Blog

Present your video blog before the class. Then answer the questions below.

Use an example to explain how relative-age dating can be used to order events in Earth's past, even though the events happened long before written records existed.

Which type of evidence is most helpful for determining the relative ages of major events in Earth's 4.6-billion-year-old history? Explain your answer.

The geologic time scale is a type of model used to study Earth. How could you use the presentations from the groups in your class to create a geologic time line?

Congratulations! You've completed the Science Challenge requirements!

Module Wrap-Up

REVISIT
THE PHENOMENON

Using the concepts you learned in the module, explain how rocks, and the fossils within, can tell the story of Earth's long history.

INQUIRY

If you had to ask one question about what you studied, what would it be?

Plan and conduct an investigation to answer this question.

Natural Selection and Adaptations

ENCOUNTER THE PHENOMENON

Why does this butterfly, when it stretches its wings, look like the face of an owl?

Whooo are you—owl or butterfly?

GO ONLINE

Watch the video *Whooo are you—owl or butterfly?* to see this phenomenon in action.

Communicate Think about why this butterfly resembles an owl. Record your ideas for why and how this occurs below. Discuss your ideas with three different partners. Revise or update your ideas, if necessary, after the discussions with your classmates.

STEM Module Project Launch
Science Challenge

Population Probabilities

The education director of an environmental studies lab has asked you to help her design and test an experiment. The experiment will combine math and science to explain how natural selection changes the frequency of a trait in a population over time.

Your task is to design and perform the experiment and then to construct an explanation of the resulting data. The education director has explained that the experiment will involve deer mice, which can be either brown or gray.

Start Thinking About It

Notice that the color of the mouse in the photo above enables it to blend in with its environment, much like the owl butterfly you just discussed. How might this help the mouse survive? Discuss your thoughts with your group.

Lesson 1
How Traits Change

Lesson 2
The Theory of Evolution by Natural Selection

Lesson 3
Artificial Selection

STEM Module Project

Planning and Completing the Science Challenge
How will you meet this goal? The concepts you will learn throughout this module will help you plan and complete the Science Challenge. Just follow the prompts at the end of each lesson!

SCIENCE PROBES

LESSON 1 LAUNCH

Cat Eyes

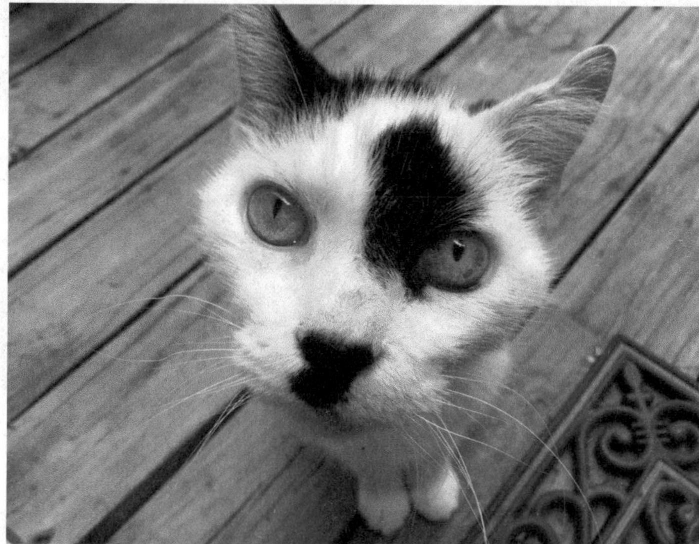

Logan and two of his friends were playing with his pet cat when he pointed out that the cat had two different colored eyes. Each of the three friends had different ideas about how this could happen. Here are their thoughts:

Logan: I think my cat has different colored eyes because he inherited the color of one eye from his mom and one from his dad.

Camila: I think the cat has different colored eyes because there is a mistake in his DNA.

Min: I think the cat has different colored eyes because his mom and dad have two different colored eyes.

Circle the person you most agree with. Explain why you agree with that person.

You will revisit your response to the Science Probe at the end of the lesson.

LESSON 1

How Traits Change

ENCOUNTER THE PHENOMENON | Why is this alligator white and not green?

Think about an alligator. You're probably picturing an animal that is green. What can happen when an alligator is a different color?

1. Start with 2 pieces of plain white and plain green paper each.
2. Take one sheet of white paper and use a paper punch to punch out 20 holes. Do the same with one sheet of green paper. Mix all the dots together in one container.
3. Place the remaining sheet of green paper on the ground. This will be your habitat. Spread the dots across the habitat.
4. Now acting as a predator, see how many dots you can collect from the habitat in 15 seconds. Record the number of green and white dots collected.

   Green ______________ *White* ______________
5. Return all of the dots to the container. Repeat steps 3 and 4, this time using the white paper as your habitat. Record your results.

   Green ______________ *White* ______________
6. Which dots were easier to collect on each habitat? Why?

7. How does this relate to the alligator in the photo?

A white alligator?

GO ONLINE Watch the video *A white alligator?* to see this phenomenon in action.

EXPLAIN
THE PHENOMENON

The alligator in the photo is white instead of green. Its physical traits have changed, giving it characteristics different from most alligators. Now that you have seen changing traits, make a claim about what causes them.

CLAIM

A change in traits may be caused by...

COLLECT EVIDENCE as you work through the lesson. Then return to these pages to record your evidence.

EVIDENCE

A. What evidence have you discovered to explain how DNA controls traits such as the color of the alligator?

B. What evidence have you discovered to explain how mutations, such as the white coloring of the alligator, occur?

MORE EVIDENCE

C. What evidence have you discovered to explain the results of mutations, such as the one that caused the white coloring of the alligator?

When you are finished with the lesson, review your evidence. If necessary, based on the evidence, revise your claim.

REVISED CLAIM

A change in traits may be caused by...

Finally, explain your reasoning for how and why your evidence supports your claim.

REASONING

The evidence I collected supports my claim because...

How does DNA affect traits?

An organism's body cells use codes to determine genetic traits. Interpret the code below to learn more about this.

INVESTIGATION

Crack the Code

1. Analyze the pattern of the simple code shown below.

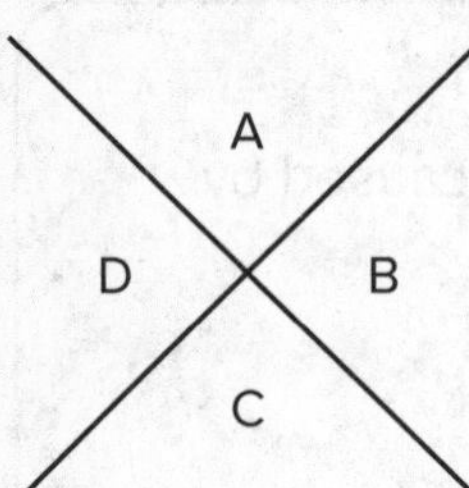

E	F	G
H	I	J
K	L	M

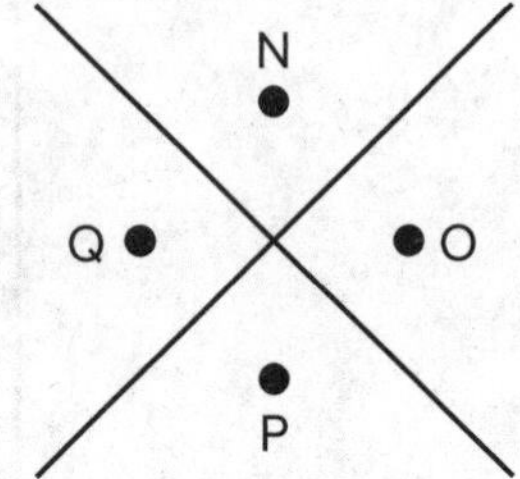

R•	S•	T•
U•	V•	W•
X•	Y•	Z•

2. Record the correct letters for the symbols in the code on the lines below.

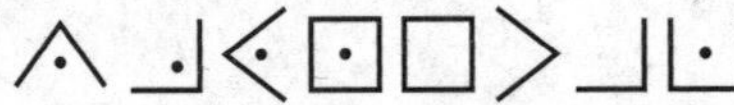 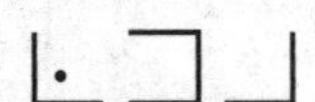 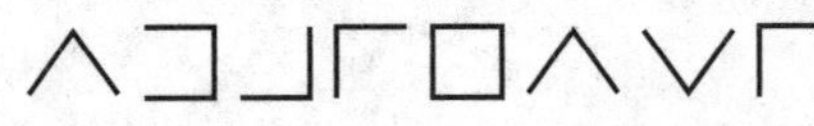

 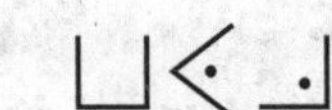  

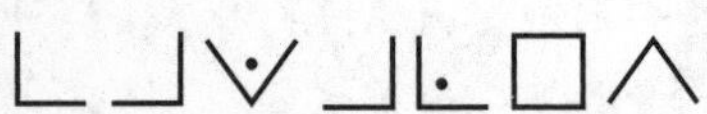 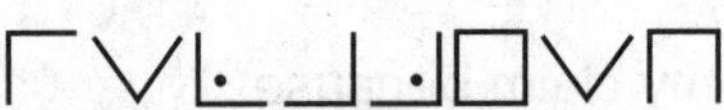

3. What do all codes, such as Morse code, Braille, and this one, have in common?

4. What do you think might happen if there is a mistake in the code?

5. How do you think an organism's cells might use code to determine its traits?

The Structure of DNA Cells put molecules together by following codes, or a set of directions. Where do those directions come from? Genes provide directions for a cell to assemble molecules that express traits. You might recall that a gene is a section of a chromosome. Chromosomes are made of proteins and **DNA**—an organism's genetic material. A gene is a segment of DNA on a chromosome.

Countless numbers of directions are needed to make all the molecules in cells and organisms. How do all these directions fit on a few chromosomes? Strands of DNA in a chromosome, shown below, are tightly coiled, like a telephone cord or a coiled spring. This coiling allows more genes to fit in a small space.

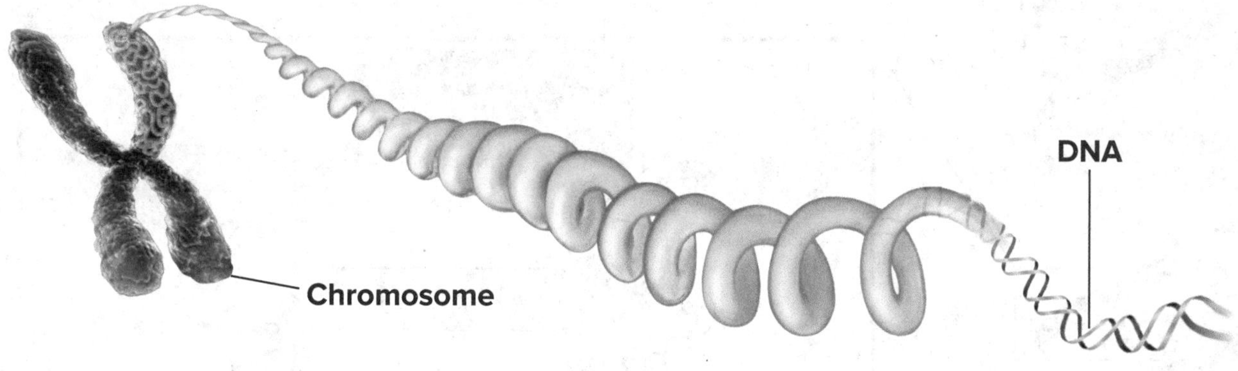

Want more information?
Go online to read more about DNA and mutations.

FOLDABLES
Go to the Foldables library to make a Foldable that will help you take notes while reading this lesson.

A Complex Molecule DNA is like a twisted zipper. This twisted zipper shape is called a double helix. A model of DNA's double helix structure is shown below.

HISTORY Connection How did scientists discover the shape of DNA? Rosalind Franklin and Maurice Wilkins were two scientists in London who used X-rays to study DNA. Some of the X-ray data indicated that DNA has a helix shape.

American scientist James Watson visited Franklin and Wilkins and saw one of the X-rays of DNA. Watson realized that the X-ray gave valuable clues about the structure of DNA. Watson worked with an English scientist, Francis Crick, to build a model of DNA.

Watson and Crick based their work on information from Franklin's and Wilkins's X-rays. They also used chemical information about DNA discovered by another scientist, Erwin Chargaff. After several tries, Watson and Crick built a model that showed how the smaller molecules of DNA bond together and form a double helix.

Four Nucleotides Shape DNA DNA's twisted-zipper shape is because of molecules called nucleotides. A **nucleotide** is a molecule made of a nitrogen base, a sugar, and a phosphate group. There are four nitrogen bases: adenine (A), cytosine (C), thymine (T), and guanine (G). A and T always bond together, and C and G always bond together. Examine the figure below to see how DNA is composed.

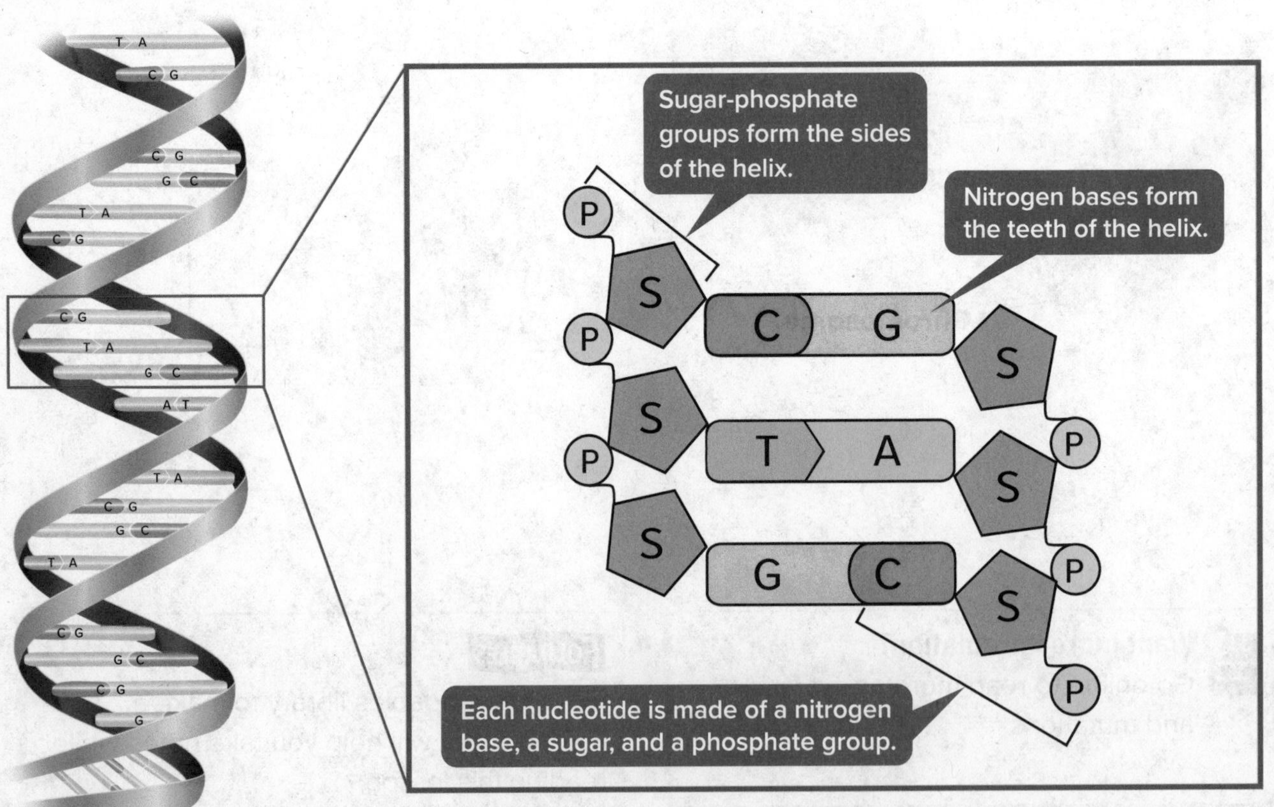
Sugar-phosphate groups form the sides of the helix.
Nitrogen bases form the teeth of the helix.
Each nucleotide is made of a nitrogen base, a sugar, and a phosphate group.
P
S
C
G
T
A
G
C

Modeling DNA The information, or directions, needed for an organism to grow, maintain itself, and reproduce is stored in DNA. Its unique structures allow this information to be kept in an extremely small area. The double helix is now a well-known molecular structure that you have probably seen before. Making a model of DNA can help you understand its structure.

LAB Model DNA

Safety 

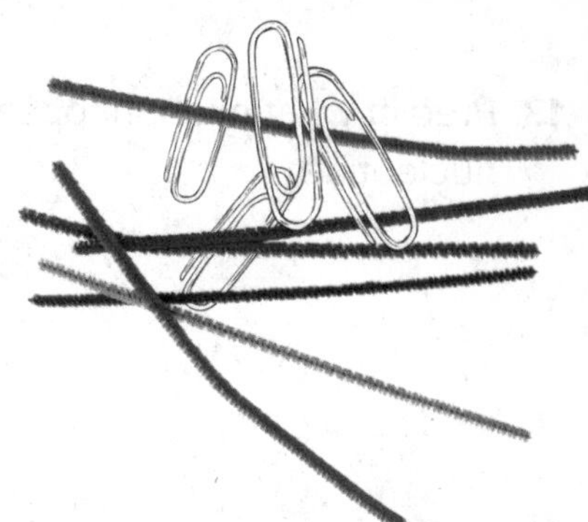

Materials

- small paper clips (10)
- large paper clips (10)
- chenille stems, four colors (10)
- styrene foam blocks (2)

Procedure

1. Read and complete a lab safety form.
2. Link a small paper clip to a large paper clip. Continue until you have made a chain of 10 paper clips.
3. Choose four colors of chenille stems. Each color represents one of the four nitrogen bases. Record the color of each nitrogen base in the space to the right.
4. Attach a chenille stem to each large paper clip.
5. Repeat step 2 and step 4, but this time attach the corresponding chenille stem nitrogen bases.
6. Connect the nitrogen bases.
7. Securely insert one end of your double chain into a block of styrene foam.
8. Repeat step 7 with the other end of your chain.
9. Gently turn the blocks to form a double helix.
10. Draw a model of your structure to the right.
11. Follow your teacher's instructions for proper cleanup.

Analyze and Conclude

12. Explain which part of a DNA molecule is represented by each material you used in your model.

13. Predict what might happen if a mistake were made in creating a nucleotide.

14. How did making a model of DNA help you understand its structure and function?

15. Why do you think a tightly coiled DNA shape is more efficient for storing information than if the shape were more like a straight line?

How DNA Replicates Cells contain DNA in chromosomes. So, every time a cell divides, all chromosomes must be copied for the new cell. The new DNA is identical to existing DNA. The process of copying a DNA molecule to make another DNA molecule is called **replication.**

GO ONLINE Watch the animation *DNA Replication* to see this process in action. After you watch the animation, follow the steps of DNA replication below.

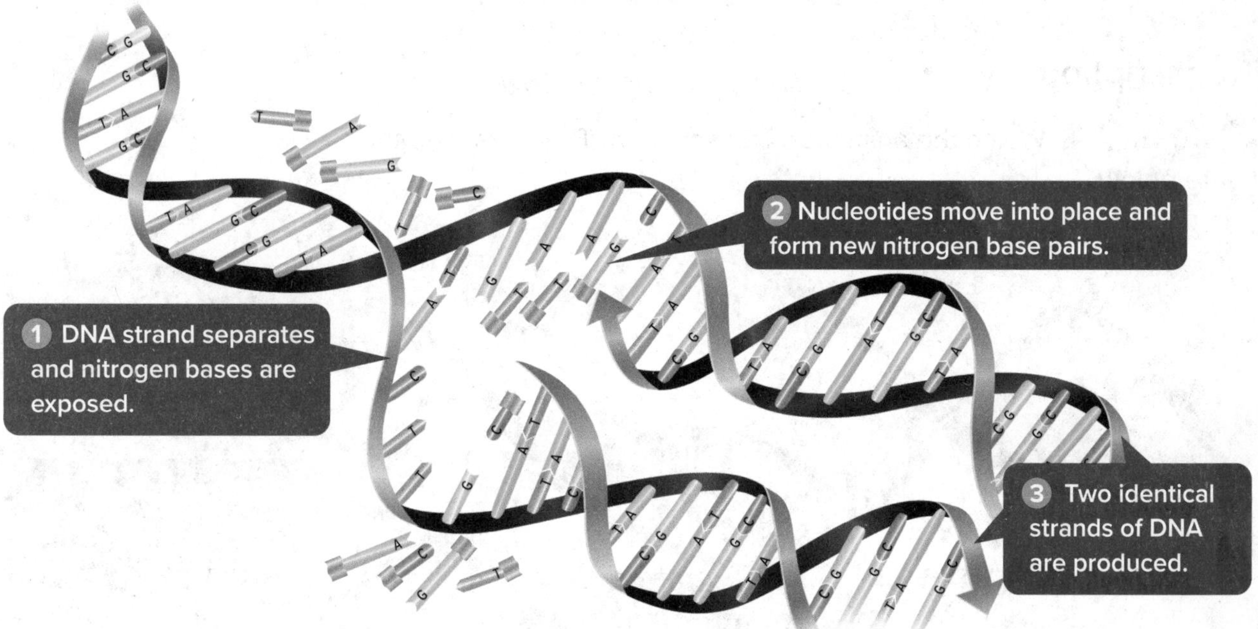

1 DNA strand separates and nitrogen bases are exposed.
2 Nucleotides move into place and form new nitrogen base pairs.
3 Two identical strands of DNA are produced.

Making Proteins Proteins are important for every cellular process. The DNA of each cell carries a complete set of genes that provides instructions for making all the proteins a cell requires. Most genes contain instructions for making proteins. Some genes contain instructions for when and how quickly proteins are made.

The Role of RNA in Making Proteins How does a cell use the instructions in a gene to make proteins? Proteins are made with the help of ribonucleic acid (**RNA**)—a type of nucleic acid that carries the code for making proteins from the nucleus to the cytoplasm. RNA also carries amino acids around inside a cell and forms a part of ribosomes.

THREE-DIMENSIONAL THINKING

Explain how the double helix **structure** of DNA aids in the **function** of replication.

Transcription RNA, like DNA, is made of nucleotides. However, there are key differences between DNA and RNA. DNA is double-stranded, but RNA is single-stranded. RNA has the nitrogen base uracil (U) instead of thymine (T) and the sugar ribose instead of deoxyribose. The first step in making a protein is to make mRNA from DNA. The process of making mRNA from DNA is called **transcription.** How does this happen? Let's find out.

INVESTIGATION

Transcription

GO ONLINE Watch the animation *Transcription*. Then examine the figure below.

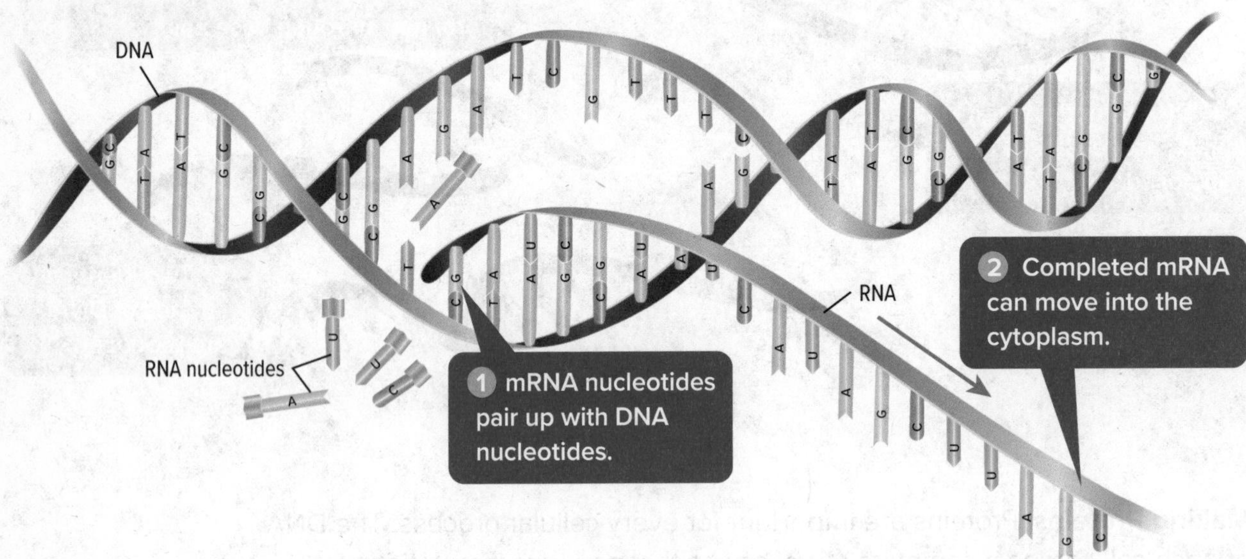

Based on the animation and the figure, complete the DNA to mRNA transcription below. Use the blue lines to write your answer.

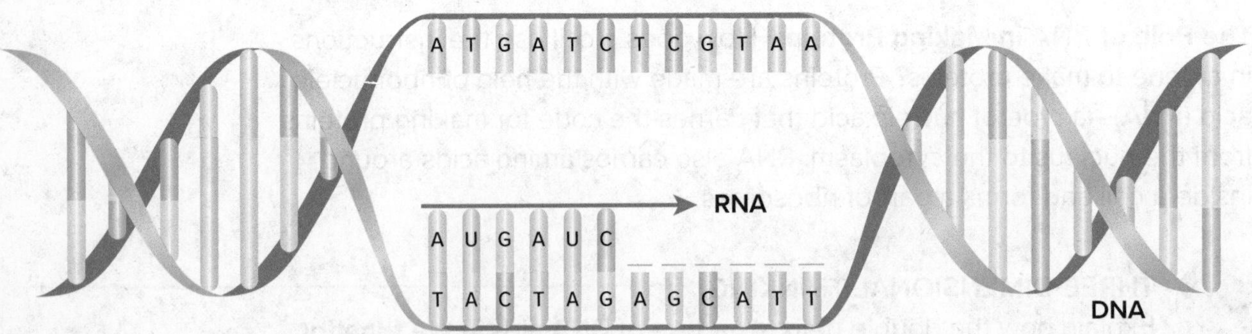

Types of RNA You just read about messenger RNA (mRNA). There are two other types of RNA, transfer RNA (tRNA) and ribosomal RNA (rRNA). The three types of RNA work together to make proteins. The process of making a protein from RNA is called **translation.** Translation occurs in ribosomes, cell organelles that are attached to the rough endoplasmic reticulum (rough ER), as shown below. Ribosomes are also in a cell's cytoplasm.

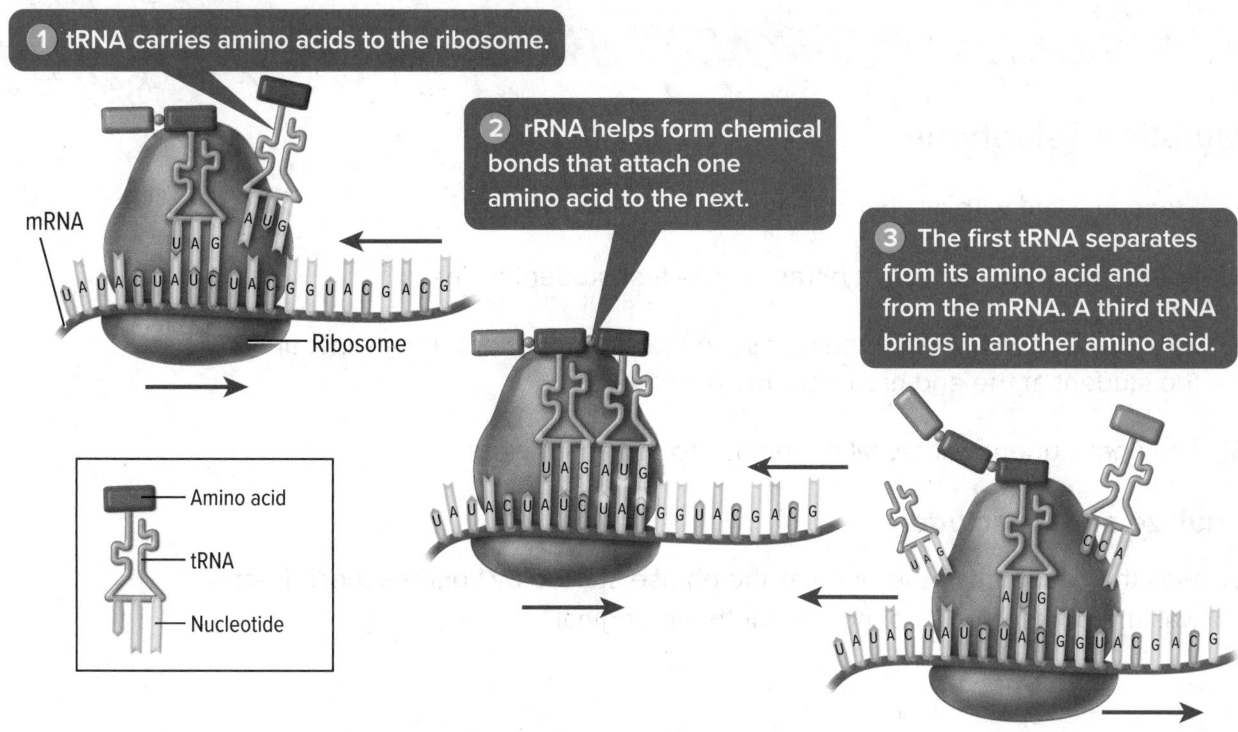

1 tRNA carries amino acids to the ribosome.
2 rRNA helps form chemical bonds that attach one amino acid to the next.
3 The first tRNA separates from its amino acid and from the mRNA. A third tRNA brings in another amino acid.
mRNA
Ribosome
Amino acid
tRNA
Nucleotide

Translating the RNA Code Making a protein from mRNA is like using a secret code. Proteins are made of amino acids. The order of the nitrogen bases in mRNA determines the order of the amino acids in a protein. Three nitrogen bases on mRNA form the code for one amino acid. Each series of three nitrogen bases on mRNA is called a codon.

THREE-DIMENSIONAL THINKING

What would be the **effect** on the **structure and function** of a cell if it were unable to produce mRNA?

COLLECT EVIDENCE

How does DNA control traits such as the color of the alligator at the beginning of the lesson? Record your evidence (A) in the chart at the beginning of the lesson.

How can a trait change?

Sometimes, mistakes can happen during replication. Most mistakes are corrected before replication is completed. A mistake that is not corrected can result in a mutation.

INVESTIGATION

ROFL
IDK
LOL

Mutation Telephone

1. Stand in a line with all your classmates.
2. Your teacher will whisper a phrase to the first student in line.
3. Each student will whisper the phrase to the student after them in line, until the student at the end has heard the phrase.
4. The final student will reveal the phrase to the entire class.

Analyze and Conclude

5. Was the final phrase the same as the phrase started by your teacher? If not, was the phrase similar or not similar to the original?

6. If the phrase changed, why do you think this occurred?

7. How do you think this could represent a model of a mutation in DNA?

Mutations Recall that genes are located on chromosomes and that each pair of chromosomes contains two variants of a gene. Each gene controls the production of specific proteins, which in turn control traits.

A **mutation** is a permanent change in the sequence of DNA. Changes in genetic material can result in the production of different proteins. This change can alter an organism's traits. Mutations can also be triggered by environmental factors such as exposure to X-rays, ultraviolet light, radioactive materials, and some chemicals.

When an organism's phenotype, how a trait appears or is expressed, changes in response to its environment, the organism's genes are not affected and the change cannot be passed on to the next generation. The only way that a trait can change so that it can be passed to the next generation is by mutation, or changing an organism's genotype.

Although all genes can mutate, only mutated genes in egg or sperm cells are inherited. Some mutations in egg or sperm cells occur if an organism is exposed to harsh chemicals or severe radiation. But most mutations occur randomly.

Types of Mutations There are many types of DNA mutations. In a deletion mutation, one or more nitrogen bases are left out of the DNA sequence. In an insertion mutation, one or more nitrogen bases are added to the DNA. In a substitution mutation, the nitrogen base is replaced by a different nitrogen base.

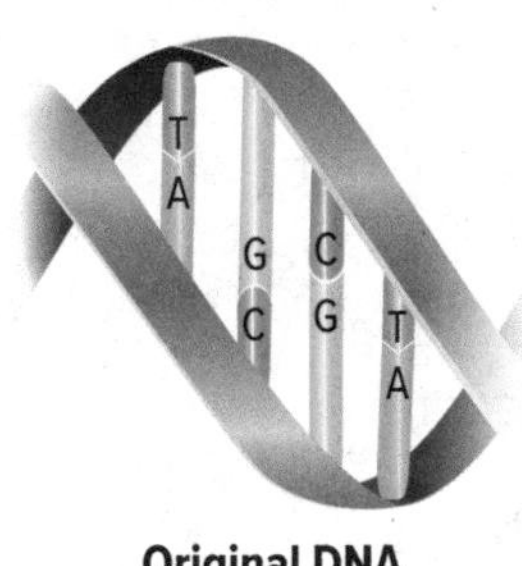

Original DNA sequence

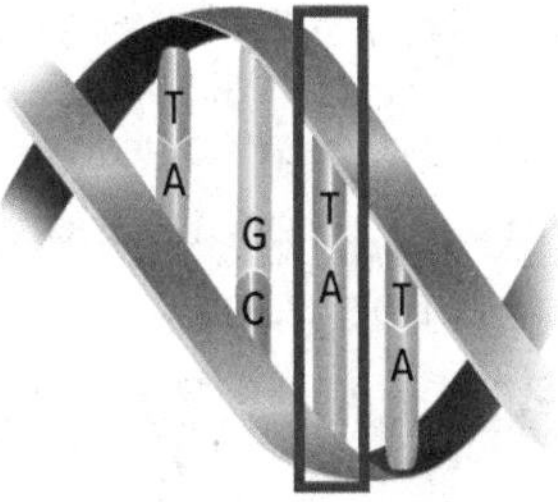

Substitution
The C-G base pair has been replaced with a T-A pair.

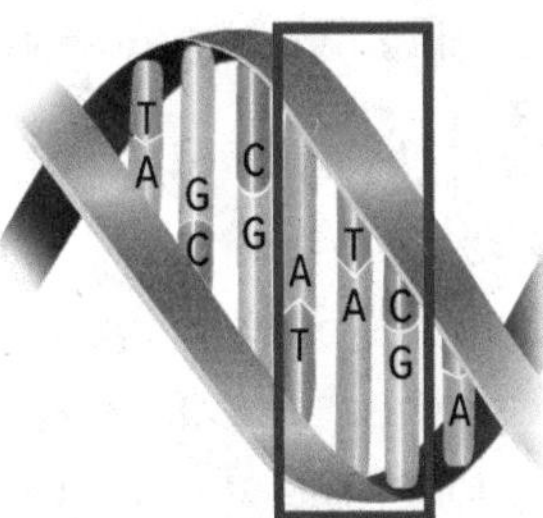

Insertion
Three base pairs have been added.

Deletion
Three base pairs have been removed. Other base pairs will move in to take their place.

COLLECT EVIDENCE

How do mutations, such as the white coloring of the alligator at the beginning of the lesson, occur? Record your evidence (B) in the chart at the beginning of the lesson.

A Closer Look: Why so blue?

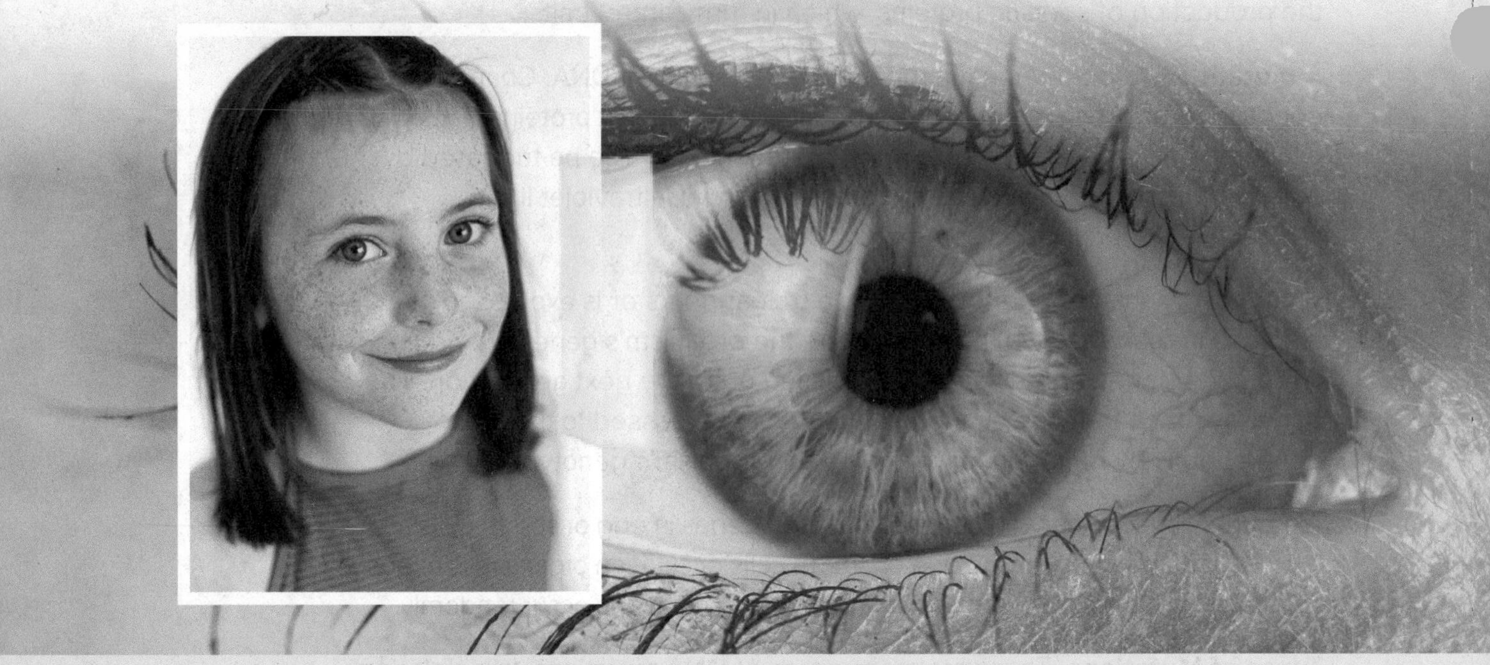

Eye color is a trait determined by the pigmentation of the eye's iris, which can come in a variety of colors, including brown, green, and hazel. You may have wondered what causes some of these colors to appear. For example, you, or someone you know, may have eyes that are blue. You may be surprised to hear that the appearance of blue eyes is actually a mutation!

The mutation that causes blue eyes has been around for about 6,000–10,000 years. Scientists have discovered a Stone Age man who lived about 7,000 years ago whose DNA revealed the code for blue eyes. He is the earliest known person to have the mutation. Scientists hypothesize that all blue-eyed individuals alive today come from a single blue-eyed ancestor!

Originally, all humans had brown eyes. However, a genetic mutation affecting the OCA2 gene in human chromosomes resulted in the turning off of the production of brown eyes. The mutation does not turn off the OCA2 gene all together. It causes a reduction in the production of melanin, or pigment, in the iris. This causes brown to diminish to blue. If the OCA2 gene were to be completely turned off, it would cause albinism which is a lack of melanin in the eyes, skin, and hair.

Blue eyes represent a mutation that is neither positive nor negative because it does not impact a person's chance of survival.

It's Your Turn

Explain Write a caption for a social media post featuring a picture of someone with blue eyes. Explain the origin of blue eyes in the post.

What happens when a mutation occurs?

The effect of a mutation depends on where in the DNA sequence the mutation happens and the type of mutation. Proteins express traits. Because mutations can change proteins, they can cause changes in traits. Some mutations might cause a trait to change in a way that benefits the organism, while others cause genetic disorders. Examine the examples below.

Due to a genetic mutation, about 1 in 5 million lobsters are blue!

Genetic Disorders		
Defective Gene or Chromosome	**Disorder**	**Description**
Chromosome 7, CFTR gene	Cystic fibrosis	In people with defective CFTR genes, salt cannot move in and out of cells normally. Mucus builds up outside cells. The mucus can block airways in lungs and affect digestion.
Chromosome 17, BRCA1; Chromosome 13, BRCA2	Breast cancer and ovarian cancer	A defect in the BRCA1 and/or BRCA2 does not mean the person will have breast cancer or ovarian cancer. People with defective BRCA1 or BRCA2 genes have an increased risk of developing breast cancer and ovarian cancer.
Chromosome 7, elastin gene	Williams Syndrome	People with Williams syndrome are missing part of chromosome 7, including the elastin gene. The protein made from the elastin gene makes blood vessels strong and stretchy.
Chromosome 12, PAH gene	Phenylketonuria (PKU)	People with defective PAH genes cannot break down the amino acid phenylalanine. If phenylalanine builds up in the blood, it poisons nerve cells.

INVESTIGATION

Mutation Classification

A mutation can be negative, positive, or neutral. A positive mutation is beneficial to an organism. A negative mutation is harmful to an organism. A neutral mutation is neither beneficial nor harmful, and may not appear in the phenotype. Classify each of the following mutations as positive, negative, or neutral. Record your thoughts in the table.

Mutation	Effects	Positive	Negative	Neutral
Eye color	Genes for brown eyes are mutated and the individual has blue eyes.			
Lactose tolerance	Due to a mutation, human adults are able to process lactose unlike other mammals.			
Color blindness	Due to a mutation on the X chromosome, a person cannot see certain colors.			

Choose one of the above mutations to research further. Finally, create a short public service announcement, in your Science Notebook, explaining one of the three mutations.

SEP DCI CCC

THREE-DIMENSIONAL THINKING

Predict the **effect** of a mutation which limits the production of pigment in hair.

COLLECT EVIDENCE

What are the results of a mutation, such as the white coloring of the alligator at the beginning of the lesson? Record your evidence (C) in the chart at the beginning of the lesson.

When Asthma Attacks

SCIENCE & SOCIETY

Is asthma caused by genes, the environment, or both?

More than 6 million children under the age of 18 in the United States have asthma. During an asthma attack, airways tighten up and become filled with mucus. This makes it hard to breathe. If you or someone in your family has asthma, you might know the signs of an asthma attack—difficulty breathing, wheezing, coughing, and tightness of the chest. While the symptoms can come and go, the disease is always there.

Asthma is caused by a combination of genes and environmental factors. Children are more likely to have asthma if a parent or close relative has it. Allergies and some respiratory infections can cause asthma. If asthma runs in a person's family, exposure to pollen, dust mites, animal hair, cockroaches, and cigarette smoke can trigger the development of asthma.

Currently, there is no cure for asthma, but medications can help people prevent or treat asthma attacks. Scientists hope to learn more about the genetic and the environmental causes of asthma so they can improve medicines and someday find a cure.

To learn more about the genetic causes of asthma, scientists studied a group of people called the Hutterites. The Hutterites live in small, isolated communities in rural parts of the north-central United States and central Canada. The Hutterites all share the same diet and lifestyle, but only some individuals have asthma. Researchers learned that Hutterites with asthma have a tiny mutation on particular genes. Researchers suspect this mutation makes Hutterites more likely to have or develop asthma. This discovery might help doctors predict and prevent asthma in Hutterites and people worldwide.

▲ **Scientists found a mutation on chromosome 1 that could be among the genetic causes in the development of asthma.**

Scientists studied the genes of the Hutterites to learn more about the causes of asthma. ▶

It's Your Turn

Write a paragraph explaining why researchers chose the Hutterites for their study.

LESSON 1

Review

Summarize It!

1. **Identify** one environmental factor and one nonenvironmental factor that can trigger a mutation. Describe how the environmental factor could be prevented.

Three-Dimensional Thinking

Use the diagram below to answer the following questions.

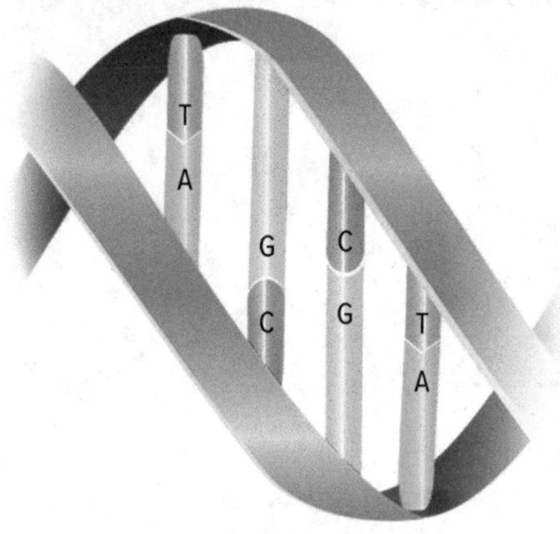

Before Replication

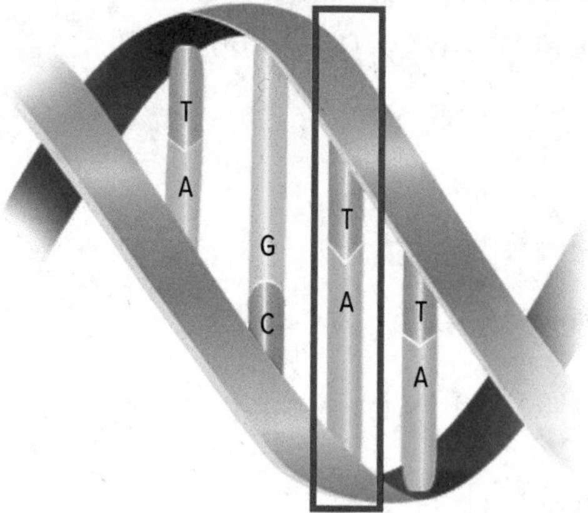

After Replication

2. The diagram above shows a segment of DNA before and after replication. Which occurred to the structure of the DNA during replication?

A deletion

C substitution

B insertion

D translation

3. The mutation shown above resulted in muscle degeneration. The effect of this mutation is that muscles become progressively weaker. What type of mutation is this?

A positive

C neutral

B negative

D none of the above

Real-World Connection

4. **Write** Your English teacher has asked you to write a short story about a superhero with a mutation that causes powers, using a real factor that causes mutations. Identify your character below, and describe the cause and effects of the mutation.

Still have questions?
Go online to check your understanding of DNA and mutations.

REVISIT

Do you still agree with the person you chose at the beginning of the lesson? Return to the Science Probe at the beginning of the lesson. Explain why you agree or disagree with that person now.

EXPLAIN THE PHENOMENON

Revisit your claim about how traits change. Review the evidence you collected. Explain how your evidence supports your claim.

START PLANNING

STEM Module Project
Science Challenge

Now that you've learned about how traits change, go to your Module Project and start planning your activity. Keep in mind that you'll want to include information about how mutations cause changes to phenotypes of organisms, such as the owl butterfly.

PAGE KEELEY SCIENCE PROBES

LESSON 2 LAUNCH

Tree Snails

A population of a tree snail species has different patterns in their shells. These slight differences in appearance among the individual members of the species are called variations. How do you think these variations get passed on from one generation of tree snails to the next? Circle the answer that best matches your thinking.

A. from parents to offspring

B. through the environment

C. from both parents to offspring and through the environment

Explain your thinking. Describe how variations are passed on from one generation to the next.

____

____

____

____

____

____

____

You will revisit your response to the Science Probe at the end of the lesson.

LESSON 2

The Theory of Evolution by Natural Selection

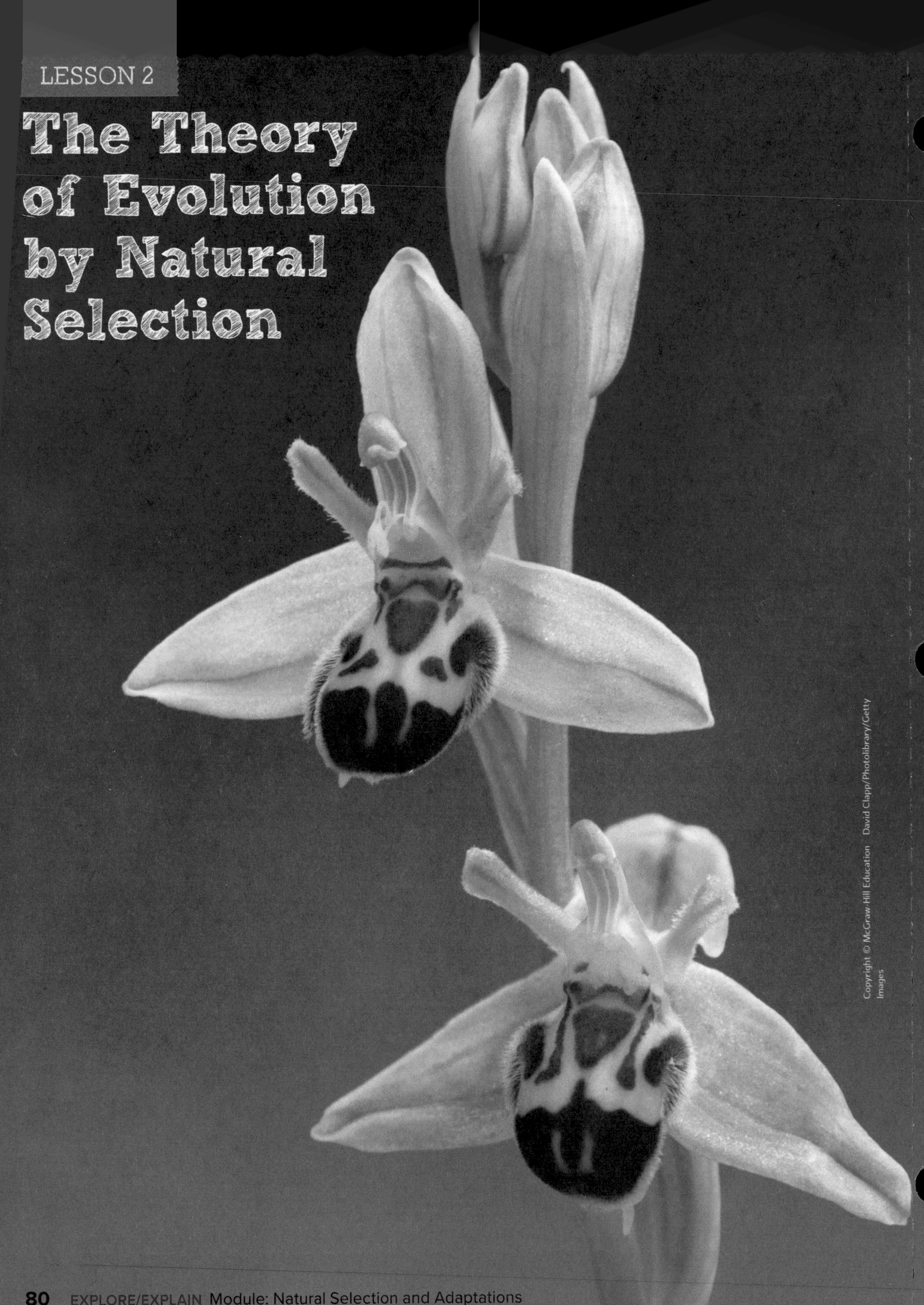

ENCOUNTER THE PHENOMENON | Why do you think this plant has flowers that look like bees?

A type of orchid plant, called a bee orchid, produces the flowers seen in the photo. You may have noticed that the flowers look like bees. Research other instances of organisms that look like a different organism. Present the organisms you discovered to your class.

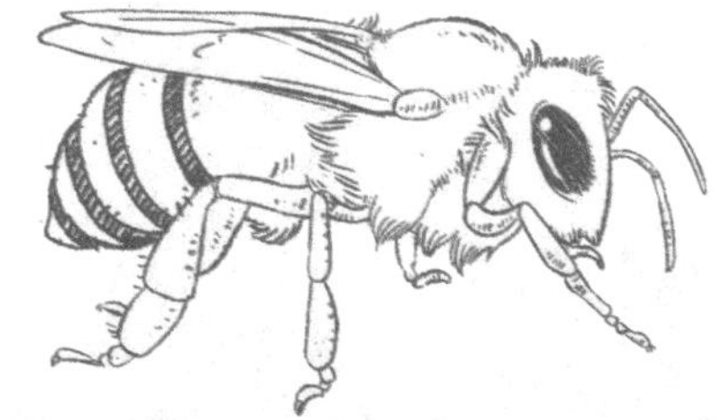

GO ONLINE
Nature's Copycats
Check out *Nature's Copycats* to see this phenomenon in action.

EXPLAIN THE PHENOMENON

Were you able to see how the bee orchid got its name? Use your observations about the phenomenon to make a claim about how organisms can change over time.

CLAIM

Organisms change over time...

COLLECT EVIDENCE as you work through the lesson. Then return to these pages to record your evidence.

EVIDENCE

A. What evidence have you discovered to explain how variations affect organisms, such as orchid plants?

B. What evidence have you discovered to explain how natural selection leads to changes in organisms, such as orchid plants?

MORE EVIDENCE

C. What evidence have you discovered to explain how adaptations affect organisms, such as orchid plants?

When you are finished with the lesson, review your evidence. If necessary, based on the evidence, revise your claim.

REVISED CLAIM

Organisms change over time...

Finally, explain your reasoning for how and why your evidence supports your claim.

REASONING

The evidence I collected supports my claim because...

What are variations?

All populations contain differences, or variations, in some characteristics of their members. For example, the tree snails at the beginning of the lesson were different sizes, different colors, and had different patterns on their shells. Variation in organisms occurs as a result of sexual reproduction and genetic mutations. Can you observe variations among your classmates?

Want more information?

Go online to read more about the theory of evolution by natural selection.

Go to the Foldables library to make a Foldable that will help you take notes while reading this lesson.

Classroom Variations

Safety 

Materials

meter stick

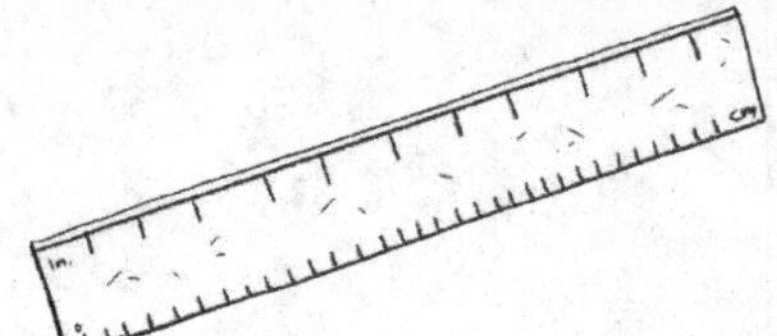

Procedure

1. Read and complete a lab safety form.
2. Use a meter stick to measure the length from your elbow to the tip of your middle finger in centimeters. Record the measurement in your Science Notebook.
3. Add your measurement to the class list.
4. Organize all the measurements from shortest to longest.
5. Break the data into regular increments, such as 31–35 cm, 36–40 cm, and 41–45 cm. Count the number of measurements within each increment. Construct a bar graph in your Science Notebook using the data. Label each axis and give your graph a title.
6. Follow your teacher's instructions for proper cleanup.

Analyze and Conclude

7. What are the shortest and longest measurements?

____

8. How much do the shortest and longest lengths vary from each other?

____

9. Describe how your results provide evidence of variations within your classroom population.

10. How do you think these variations affect you and your classmates?

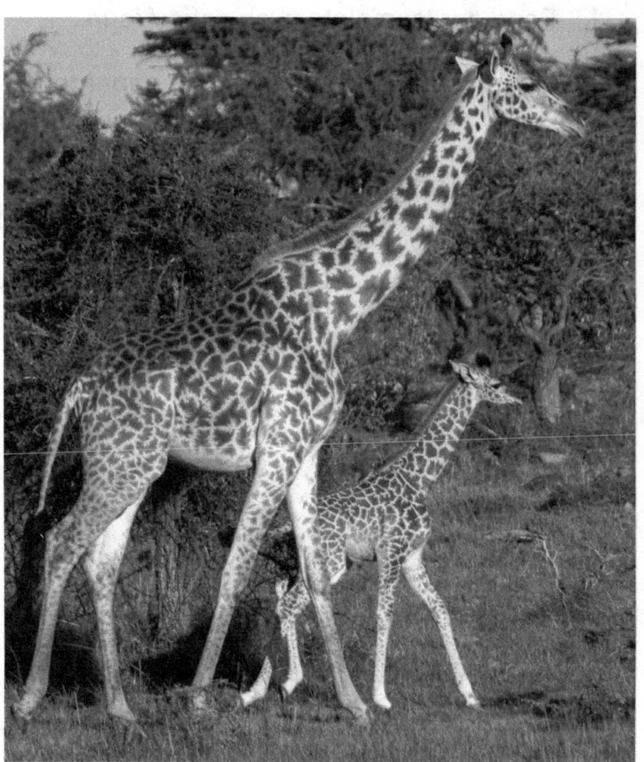

Look at the differences in the spots!

Variation In all species that reproduce sexually, offspring are different from their parents. The giraffes on the right are members of the same species, yet each one has a slightly different pattern of spots on its coat. Slight differences in inherited traits among individual members of a species are **variations.**

Variations occur through mutations. A mutation might harm an organism's chances of survival. However, many mutations, such as those that cause the unique pattern of spots on a giraffe, cause no harm. Still other mutations can benefit an organism. Mutations produce traits that help an organism survive. Think back to the bee orchids at the beginning of the lesson. The physical traits that make the flowers look like bees are the results of mutations. These traits help the orchids reproduce because bees think that the flowers are other bees and land on them in an attempt to mate with them. As a result, the plants are pollinated.

How do variations affect organisms?

Giraffe spots were probably the result of a mutation that occurred in an individual giraffe many generations ago. The mutation produced a variation that helped the giraffe survive. Explore more about variations in the lab.

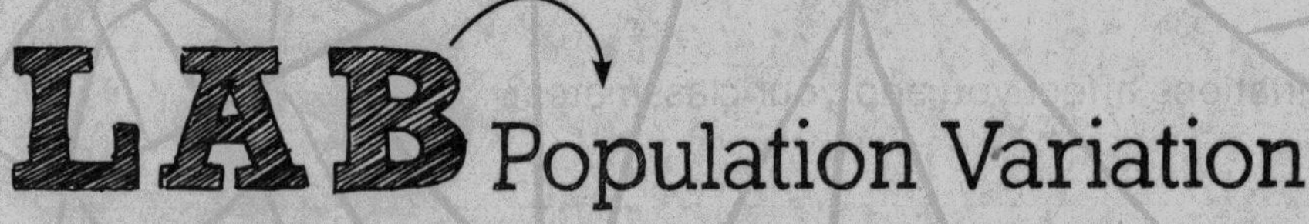

LAB Population Variation

Safety  

Materials

sunflower seeds (10)
magnifying lens
paper towel
metric ruler

Procedure

1. Read and complete a lab safety form.
2. Place 10 sunflower seeds on a paper towel. Number the seeds 1–10 by writing on the paper towel below each seed.
3. Use a magnifying lens to examine the seeds, focusing on how their coloration is alike and/or different. Record your observations below.

4. Complete the table on the next page. Perform the following steps and record your observations.
   a. Use a metric ruler to measure the length of each seed.
   b. Measure the thickness of each seed at its thickest point.

Seed Variations		
Seed	Length (mm)	Thickness (mm)
1		
2		
3		
4		
5		
6		
7		
8		
9		

5. Compare the length and thickness of your 10 seeds with other teams. Record your observations below.

6. Follow your teacher's instructions for proper cleanup.

Analyze and Conclude

7. Do all sunflower seeds have the same length and thickness? Why do you think the seeds differed in so many ways?

8. If you were a bird, do you think you would be more or less attracted to any of the seeds? How might this affect the reproduction of the sunflowers?

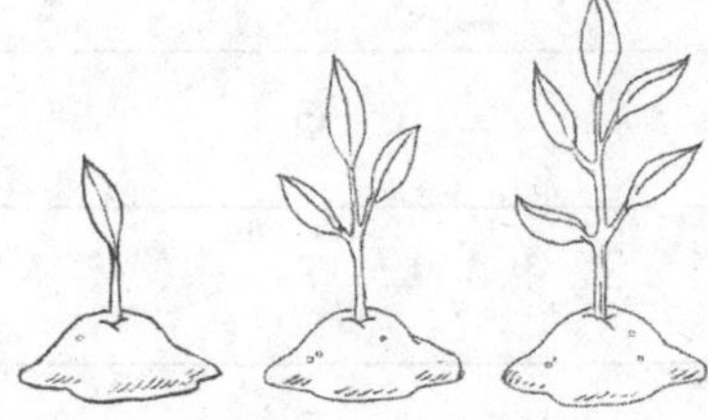

Effects of Variations Did you see how variations in the seeds could affect whether or not the sunflowers could successfully reproduce? Variations can have positive, negative, or neutral effects not only on individual organisms, but also on entire populations of organisms.

COLLECT EVIDENCE

How do variations affect organisms, such as orchid plants? Record your evidence (A) in the chart at the beginning of the lesson.

HISTORY Connection The giraffes that you saw earlier in the lesson had spots that helped them blend in with their environment—the grasslands of Africa. Recall that the spots were probably the result of a mutation that occurred in an individual giraffe. Eventually, the mutated gene became part of the giraffe population genotype. How did this happen? How can a variation in a single individual become common to an entire population? One scientist who worked to answer this question was Charles Darwin. Darwin was an English naturalist who, in the mid-1800s, developed a theory of how evolution works. A naturalist is a person who studies plants and animals by observing them. Darwin spent many years observing plants and animals in their natural habitats before developing his theory. A theory is an explanation of the natural world that is well supported by evidence.

Voyage of the Beagle Darwin served as a naturalist on the HMS Beagle, a survey ship of the British navy. During his voyage around the world, Darwin observed and collected many plants and animals. Darwin was especially interested in the organisms he saw on the Galápagos (guh LAH puh gus) Islands. The islands, shown in the figure below, are located 1,000 km off the South American coast in the Pacific Ocean. Darwin saw that each island had a slightly different environment. Some were dry. Some were more humid. Others had mixed environments.

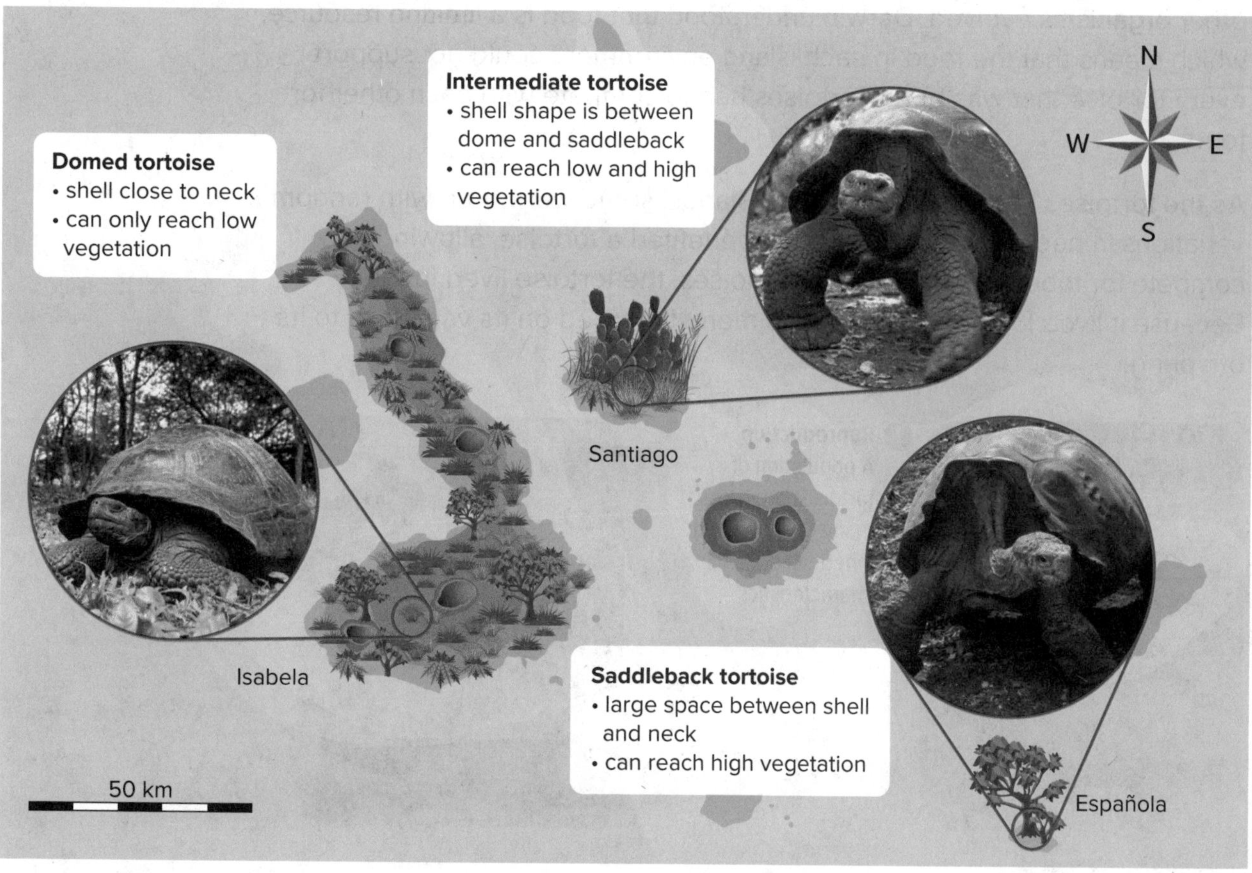

Tortoises look different depending on which island environment they inhabit.

Tortoises Giant tortoises lived on many of the islands. When a resident told Darwin that the tortoises on each island looked different, as shown in the figure above, he became curious.

Mockingbirds and Finches Darwin also became curious about the variety of mockingbirds and finches he saw and collected on the islands. Like the tortoises, different types of mockingbirds and finches lived in different island environments. Later, he was surprised to learn that many of these varieties were different enough to be separate species.

Darwin's Theory Darwin realized there was a relationship between each species and the food sources of the island it lived on. He noticed that tortoises with long necks lived on islands that had tall cacti. Their long necks enabled them to reach high to eat the cacti. The tortoises with short necks lived on islands that had plenty of short grass.

Common Ancestors Darwin became convinced that all the tortoise species were related. He thought they all shared a common ancestor. He suspected that a storm had carried a small ancestral tortoise population to one of the islands from South America millions of years before. Eventually, the tortoises spread to the other islands. Their neck lengths and shell shapes changed to match their islands' food sources.

Natural Selection Darwin did not know about genes. But he realized that variations were the key to the puzzle of how populations of tortoises and other organisms evolved. Darwin understood that food is a limiting resource, which means that the food in each island environment could not support every tortoise that was born. Tortoises had to compete with each other for food.

As the tortoises spread to the various islands, some were born with random variations in neck length. If a variation benefited a tortoise, allowing it to compete for food better than other tortoises, the tortoise lived longer. Because it lived longer, it reproduced more. It passed on its variations to its offspring.

1 Reproduction A population of tortoises produces many offspring that inherit its characteristics.

2 Variation A tortoise is born with a variation that makes its neck slightly longer.

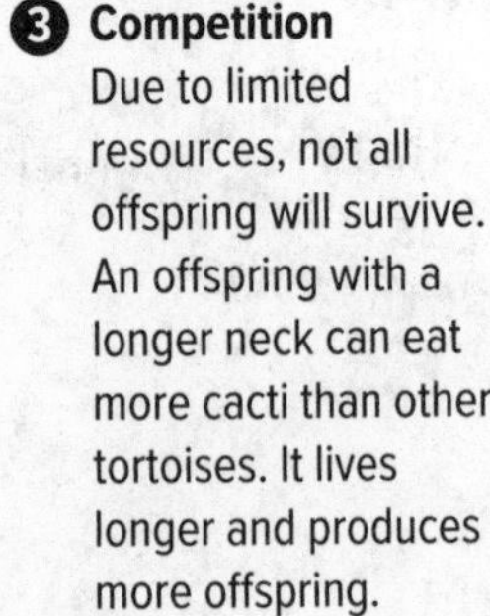

3 Competition Due to limited resources, not all offspring will survive. An offspring with a longer neck can eat more cacti than other tortoises. It lives longer and produces more offspring.

4 Selection Over time, the variation is inherited by more and more offspring. Eventually, all tortoises have longer necks.

Theory of Evolution This is Darwin's theory of evolution by natural selection. **Natural selection** is the process by which populations of organisms with variations that help them survive in their environments live longer, compete better, and reproduce more than those that do not have the variations. Natural selection explains how populations change as their environments change. It explains the process by which Galápagos tortoises became matched to their food sources. It also explains the diversity of the Galápagos finches and mockingbirds. Birds with beak variations that help them compete for food live longer and reproduce more.

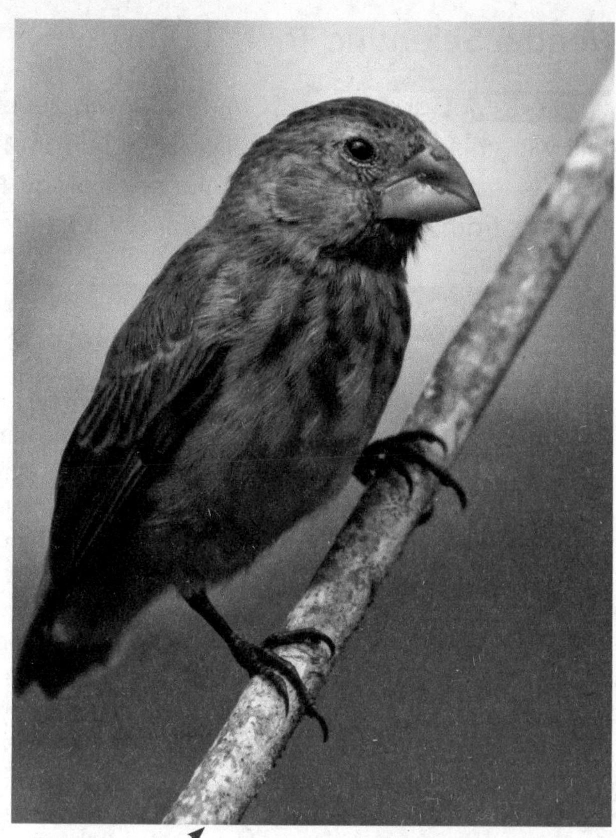

Darwin studied the diversity of Galápagos finch beaks.

COLLECT EVIDENCE

How does natural selection lead to changes in organisms, such as orchid plants? Record your evidence (B) in the chart at the beginning of the lesson.

GO ONLINE for additional opportunities to explore!

Investigate natural selection by performing one of the following activities.

☐ **Interact** with the **PhET Interactive Simulation** *Natural Selection.*

OR

☐ **Watch** the **Animation** *Natural Selection.*

THREE-DIMENSIONAL THINKING

Explain the multiple **causes** that led to the changes in the tortoises on the Galápagos Islands.

Read a Scientific Text

HISTORY Connection Charles Darwin developed the theory of evolution following his journey on the HMS Beagle from 1831–1836. Today Darwin's work is the foundation of evolutionary biology. Read an excerpt about his findings from his book *On the Origin of Species* below.

CLOSE READING

Inspect

Read the passage from *On the Origin of Species* written by Charles Darwin.

Find Evidence

Reread the passage. Then with a partner re-write the passage in your own words in your Science Notebook.

Make Connections

Communicate Pair up with another set of students and share your summaries of the text. Discuss whether or not there is anything you would like to change in your summary after hearing the other group's summary.

Natural Selection

Let it be borne in mind how infinitely complex and close-fitting are the mutual relations of all organic beings to each other and to their physical conditions of life. Can it, then, be thought improbable, seeing that variations useful to man have undoubtedly occurred, that other variations useful in some way to each being in the great and complex battle of life, should sometimes occur in the course of thousands of generations? If such do occur, can we doubt (remembering that many more individuals are born than can possibly survive) that individuals having any advantage, however slight, over others, would have the best chance of surviving and of procreating their kind? On the other hand, we may feel sure that any variation in the least degree injurious would be rigidly destroyed. This preservation of favourable variations and the rejection of injurious variations, I call Natural Selection. Variations neither useful nor injurious would not be affected by natural selection, and would be left a fluctuating element, as perhaps we see in the species called polymorphic.

We shall best understand the probable course of natural selection by taking the case of a country undergoing some physical change, for instance, of climate. The proportional numbers of its inhabitants would almost immediately undergo a change, and some species might become extinct. We may conclude, from what we have seen of the intimate and complex manner in which the inhabitants of each country are bound together, that any change in the numerical proportions of some of the inhabitants, independently of the change of climate itself, would most seriously affect many of the others. [...] Every slight modification, which in the course of ages chanced to arise, and which in any way favoured the individuals of any of the species, by better adapting them to their altered conditions, would tend to be preserved; and natural selection would thus have free scope for the work of improvement.

Source: On the Origin of Species by Means of Natural Selection, or the Preservation of Favoured Races in the Struggle for Life By Charles Darwin

Peter and Rosemary Grant

CAREERS in SCIENCE

Observing Natural Selection

Charles Darwin was a naturalist during the mid-1800s. Based on his observations of nature, he developed the theory of evolution by natural selection. Do scientists still work this way—drawing conclusions from observations? Is there information still to be learned about natural selection? The answer to both questions is yes.

Peter and Rosemary Grant are naturalists who observed finches in the Galápagos Islands for more than 30 years. They have found that variations in the finches' food supply determine which birds will survive and reproduce. They have observed natural selection in action.

The Grants live on Daphne Major, an island in Galápagos, for part of each year. They observe and take measurements to compare the size and shape of finches' beaks from year to year. They also examine the kinds of seeds and nuts available for the birds. They use this information to relate changes in the food supply to changes in the finch species' beaks.

The island's ecosystem is fragile, so the Grants take great care not to change the environment of Daphne Major as they observe the finches. They carefully plan their diet to avoid introducing new plant species to the island. They bring all the freshwater they need and they wash in the ocean. For the Grants, it's part of the job. As naturalists, they try to observe without interfering with the habitat in which they are living.

This large ground finch is one of the kinds of birds studied by the Grants.

It's Your Turn

WRITING Connection Find out more about careers in evolution, ecology, or population biology. What kind of work is done in the laboratory? What kind of work is done in the field? Write a report using evidence from informational texts to explain your findings.

What are adaptations?

Adaptations The accumulation of many similar variations can lead to an adaptation. An **adaptation** is an inherited trait that helps a species survive in its environment. Giraffes have different spot patterns, but each has spots. The spots help the giraffes blend in with their environment. As a result, predators of giraffes cannot see them as easily. The spotted coat of giraffes is an adaptation.

Types of Adaptations Every species has many adaptations. Scientists classify adaptations into three categories: structural, behavioral, and functional. Structural adaptations involve color, shape, and other physical characteristics. The shape of a tortoise's neck is a structural adaptation. Behavioral adaptations involve the way an organism behaves or acts. Hunting at night and moving in herds are examples of behavioral adaptations. Functional adaptations involve internal body systems that affect biochemistry. A drop in body temperature during hibernation is an example of a functional adaptation.

◀ **Structural Adaptation** The jackrabbit's powerful legs help it run fast to escape from predators.

▲ **Functional Adaptation** The blood vessels in the jackrabbit's ears expand to enable the blood to cool before re-entering the body.

Behavioral Adaptation▶ The jackrabbit stays still during the hottest part of the day, helping it conserve energy.

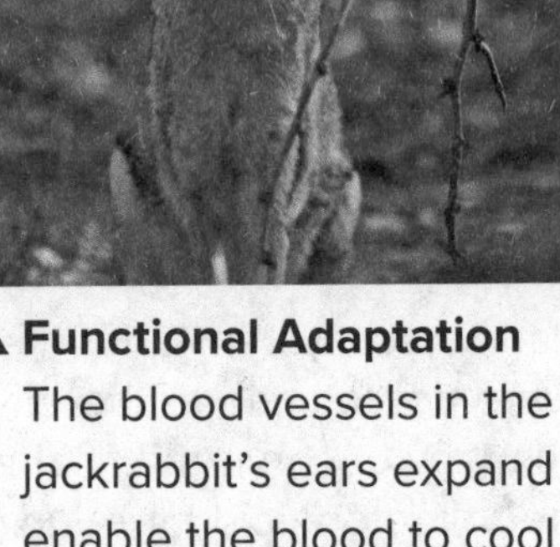

THREE-DIMENSIONAL THINKING

Primates such as humans and chimpanzees have opposable thumbs, which means that the thumb can move around to touch the other fingers on that hand. What type of adaptation is this? **Explain** why opposable thumbs are a beneficial adaptation. Record your response in your Science Notebook.

How do adaptations affect organisms?

As you have discovered, there are many different types of adaptations that help organisms survive. One type of adaptation helps organisms blend in with their environments. Can you design an organism that blends in with its environment?

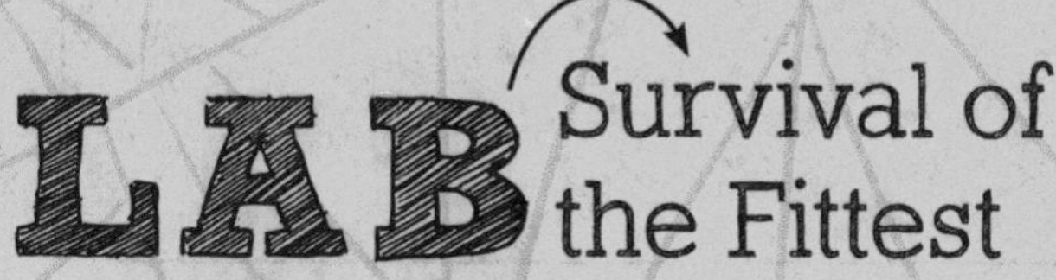

LAB Survival of the Fittest

Safety   

Materials

scissors

paper

markers

metric ruler

Procedure

1. Read and complete a lab safety form.
2. You will be designing a model of a moth. Choose an area of your classroom where your moth will rest with open wings during the day.
3. What physical traits would be beneficial to the moth based on the area selected?

   ____

   ____

   ____

4. Use scissors, paper, markers, and a ruler to design a moth that measures 2–5 cm in width with open wings and will be camouflaged where it is placed. Write the location where the moth is to be placed. Give the location and your completed moth to your teacher.
5. On the following day, you will have 1 minute to spot as many moths in the room as you can.

6. Record the location of moths spotted by your team.

7. Find the remaining moths that were not spotted. Observe their appearance.

8. Follow your teacher's instructions for proper cleanup.

Analyze and Conclude

9. Compare the appearances and resting places of the moths that were spotted with those that were not spotted.

10. Construct an explanation for how camouflage enables an organism to survive in its environment.

11. How do the results of this experiment compare with the information you have gained from the text in this lesson? Record your response in your Science Notebook.

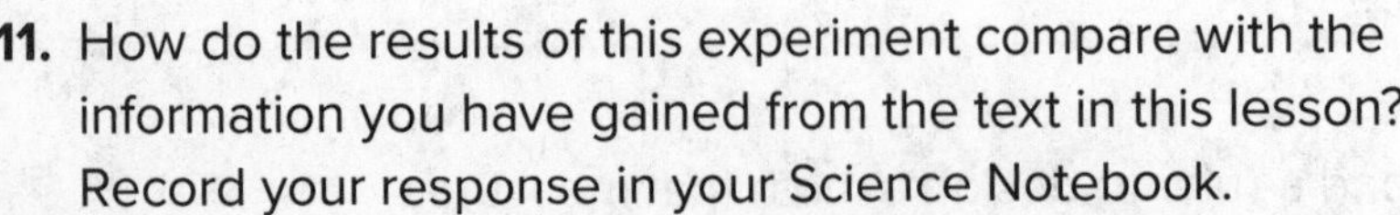

Environmental Interactions Have you ever wanted to be invisible? Many species have evolved adaptations that make them nearly invisible. The snake in the photo on the left is the same color as the leaves it is resting on. This is a structural adaptation called camouflage (KAM uh flahj). **Camouflage** is an adaptation that enables a species to blend in with its environment.

Some species have adaptations that draw attention to them. The caterpillar in the center photo resembles a snake. Predators see it and are scared away. The resemblance of one species to another species is **mimicry** (MIH mih kree). Camouflage and mimicry are adaptations that help species avoid being eaten. Many other adaptations help species eat. The pelican in the photo on the right has a beak and mouth uniquely adapted to its food source—fish.

Environments are complex. Species must adapt to an environment's living parts as well as to an environment's nonliving parts. Nonliving things include temperature, habitat, nutrients in the soil, and climate. Deciduous trees shed their leaves due to changes in climate. Camouflage, mimicry, and mouth shape are adaptations mostly to an environment's living parts.

Living and nonliving factors are always changing. Even slight environmental changes affect how species adapt. If a species is unable to adapt, it becomes extinct. The fossil record contains many fossils of species unable to adapt to change.

COLLECT EVIDENCE

How do adaptations affect organisms, such as orchid plants? Record your evidence (C) in the chart at the beginning of the lesson.

LESSON 2

Review

Summarize It!

1. Think back to your research from the Encounter the Phenomenon activity at the beginning of the lesson. Using one of the organisms that you researched, write a narrative that explains how the adaptation came about. Present your story to the class.

Three-Dimensional Thinking

Use the graph below to answer question 2.

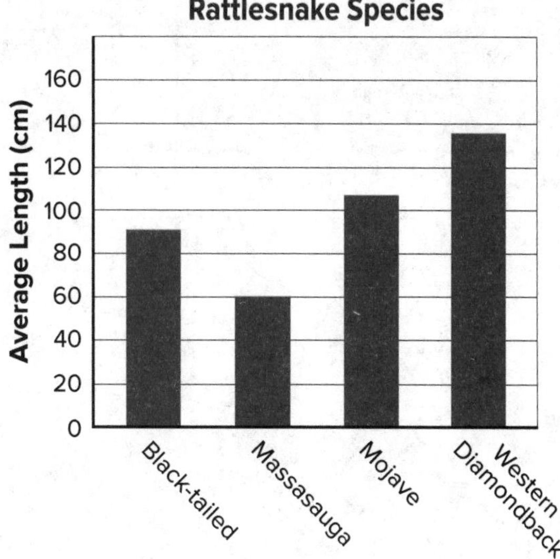

2. Which type of information about rattlesnakes does the bar graph above show?

   A adaptation

   B evolution

   C mimicry

   D variation

---

3. Which structural genetic change in the finches can be identified as the one most influenced by feeding habits, as proposed by Charles Darwin?

   A ability to fly from island to island to find the food they prefer

   B beak size and shape to take advantage of the food they had

   C claw shapes for perching on limbs while catching insects in their beaks

   D cooperative behavior so they could share limited seeds and nectar

Real-World Connection

4. **Brainstorm** Do you own clothes with a camouflage pattern? These are designed to help you blend in outdoors. What other organisms can you think of that use camouflage?

Still have questions?
Go online to check your understanding of the theory of evolution by natural selection.

REVISIT

Do you still agree with the statement you chose at the beginning of the lesson? Return to the Science Probe at the beginning of the lesson. Explain why you agree or disagree with that statement now.

EXPLAIN THE PHENOMENON

Revisit your claim about how organisms change over time. Review the evidence you collected. Explain how your evidence supports your claim.

KEEP PLANNING

STEM Module Project
Science Challenge

Now that you've learned about variations and adaptations, go back to your Module Project and continue to work on your activity. You'll want to think about how variations and adaptations can be beneficial for organisms, such as the owl butterfly.

PERFORMANCE EXPECTATION

SCIENCE PROBES

LESSON 3 LAUNCH

Corn Connection

Three friends were working on their history homework together when they noticed that the image of corn in their textbook looked a lot different than what corn looks today. Here are their thoughts:

Deidra: I think the corn from the history book is a different species than the corn we eat today.

Jayden: I think that the corn is the same, but it has changed over time.

Natalia: Even though they look different, I think the corn is the same today.

Circle the student you agree with most. Explain your choice.

____

____

____

____

____

____

You will revisit your response to the Science Probe at the end of the lesson.

LESSON 3

Artificial Selection

ENCOUNTER THE PHENOMENON | Why do all of these dogs look different?

What traits do you desire in a dog? Is there a way to get these traits? In the space below, brainstorm how you think humans can control the traits of dogs.

GO ONLINE

Dog Days

Watch the video *Dog Days* to see this phenomenon in action.

EXPLAIN
THE PHENOMENON

Did you notice the different traits in the different types of dogs? Now that you have brainstormed some ideas, use your observations about the phenomenon to make a claim about how humans can influence traits in dogs or other organisms.

CLAIM

Humans can influence traits in organisms....

COLLECT EVIDENCE as you work through the lesson.

Then return to these pages to record your evidence.

EVIDENCE

A. What evidence have you discovered to explain how humans influence traits of organisms, such as dogs, through selective breeding?

MORE EVIDENCE

B. What evidence have you discovered to explain how humans influence traits in organisms, such as dogs, through genetic engineering?

When you are finished with the lesson, review your evidence. If necessary, based on the evidence, revise your claim.

REVISED CLAIM

Humans can influence traits in organisms....

Finally, explain your reasoning for how and why your evidence supports your claim.

REASONING

The evidence I collected supports my claim because...

How can traits be directly influenced?

Adaptations provide evidence of how closely Earth's species match their environments. This is exactly what Darwin's theory of evolution by natural selection predicted. Darwin also had a hobby of breeding domestic pigeons where he found his rules of natural selection also applied in situations he controlled.

LAB Developing Dogs

Safety 

Materials

penny

colored pencils

Procedure

1. Read and complete a lab safety form.
2. In your Science Notebook, draw a dog with your desired traits. Include either long or short hair, pointed or floppy ears, light fur or dark fur, solid coloration or spotted coloration, and playful or tame. Circle your selections below:

   long hair/short hair

   pointed ears/floppy ears

   light fur/dark fur

   solid coloration/spotted coloration

   playful/tame

3. Find a partner. Imagine your dog and your partner's dog have three puppies. Choose one dog to be the mother and one to be the father.
4. Flip a coin to determine whether a puppy will inherit each of the traits from the mother (heads) or the father (tails). Repeat for all three puppies.
5. Illustrate the puppies in your Science Notebook.

6. Choose the trait you find most important in the dog. Find a partner who also chose that trait and repeat steps 3–5 with a new partner.

7. Follow your teacher's instructions for proper cleanup.

Analyze and Conclude

8. How do the puppies differ from their parents?

9. How were you able to produce a desired trait in your puppies? Were you always able to produce the desired trait? Could the traits be predicted in terms of probability?

10. How are natural selection and selective breeding related?

Want more information?
Go online to read more about artificial selection.

FOLDABLES®
Go to the Foldables® library to make a Foldable® that will help you take notes while reading this lesson.

Selective Breeding Watching natural selection in action is like watching mountains grow taller. It occurs over so many generations that it usually cannot be seen. It is easier to observe a type of selection practiced by humans. When humans breed organisms for food or for companionship, they are selecting variations that occur naturally in populations. The selection and breeding of organisms with desired traits is **selective breeding.** Selective breeding, sometimes referred to as artificial selection, is similar to natural selection except that humans, instead of nature, do the selecting. By breeding organisms with desired traits, humans change traits just as natural selection does. Cows with increased levels of milk production, dogs of different sizes, and roses of unique colors are products of selective breeding.

The fantail pigeon (top right) and the pouter pigeon (bottom right) were derived from the wild rock pigeon (above).

THREE-DIMENSIONAL THINKING

Can traits of organisms always be predicted with selective breeding? **Explain** how multiple **causes** can influence the traits of an organism.

COLLECT EVIDENCE

How can humans influence traits of organisms, such as the dogs at the beginning of the lesson, through selective breeding? Record your evidence (A) in the chart at the beginning of the lesson.

What is genetic engineering?

Recall that chromosomes are made of DNA and are in the nucleus of a cell. Sections of DNA in chromosomes that direct cell activities are called genes. Scientists are experimenting with **genetic engineering,** which are the biological and chemical methods that change the arrangement of DNA that makes up a gene. Genetic engineering already is used to help produce large volumes of medicine. Genes also can be inserted into cells to change how those cells perform their normal functions, as shown in the figure below.

Recombinant DNA Making recombinant DNA is one method of genetic engineering. Recombinant DNA is made by inserting a useful segment of DNA from one organism into a bacterium, as shown below. Large quantities of human insulin are made by some genetically engineered organisms. People with Type 1 diabetes need this insulin because their pancreases produce little to no insulin. Another use includes the production of chemicals to treat cancer.

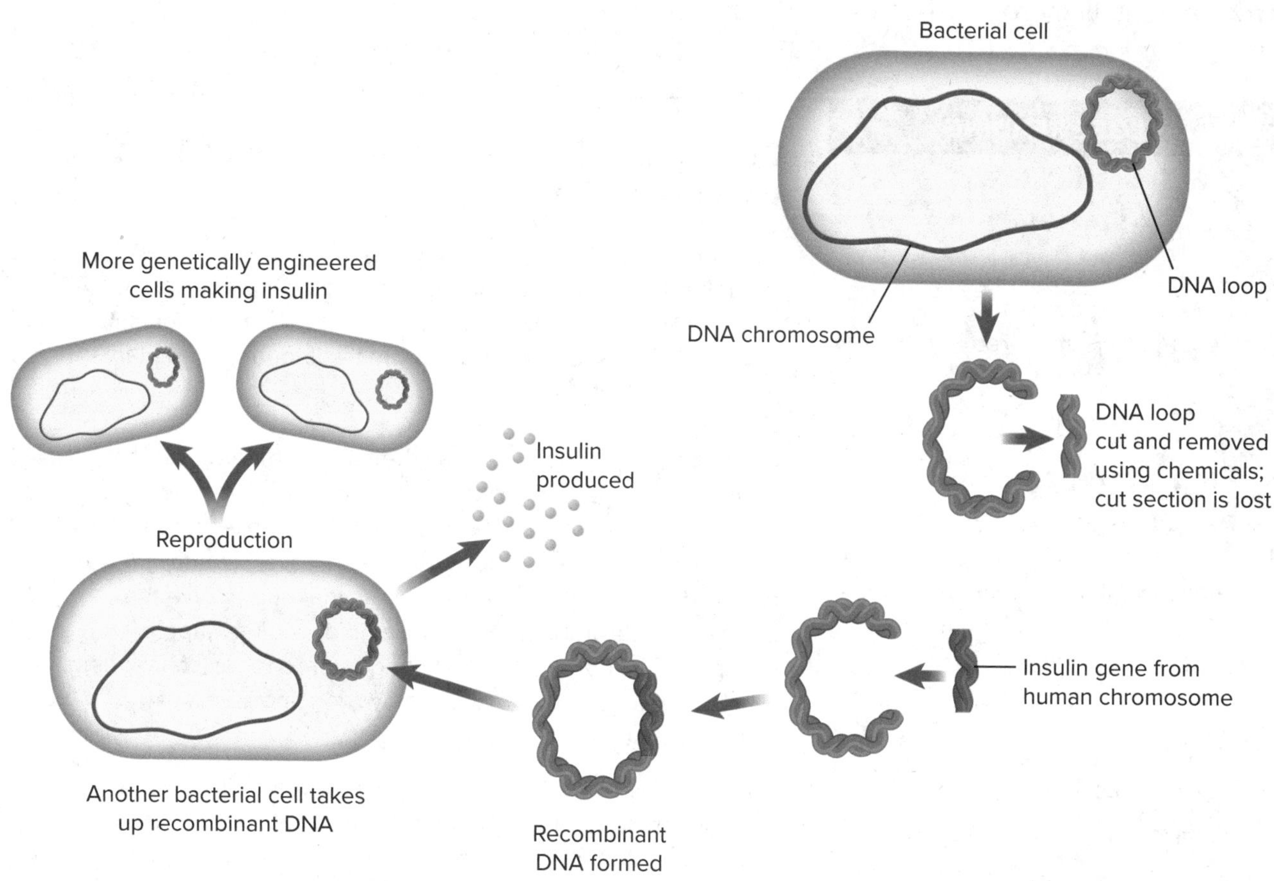

What is a genetically modified organism?

Genetic engineering can produce improvements in crop plants, such as corn, wheat, and rice. Food products that have been genetically modified are commonly referred to as genetically modified organisms, or GMOs. Scientists have made genetically engineered tomatoes with a gene that allows tomatoes to be picked while green and transported great distances before they ripen completely. Ripe, firm tomatoes are then available in the local market. Some crops are even engineered to be toxic to particular insects and pests.

Because some people might prefer foods that are not changed genetically, some stores label such items. Many people worry about possible health risks that may be present with the consumption of genetically modified crops. Others worry about the effects that altered plants might have on the environment. Is there cause to worry? Read to find out.

CLOSE READING

Inspect

Read the passage *Consumer Info About Food from Genetically Engineered Plants.*

Find Evidence

Reread the passage. Highlight the definition of genetic engineering, then underline the desirable traits resulting from genetic engineering.

Make Connections

Communicate Choose a fruit, then pair with a partner who has selected a different fruit. Design a genetically engineered fruit that could be developed from traits belonging to one of the two fruits you have chosen.

Consumer Info About Food from Genetically Engineered Plants

FDA regulates the safety of food for humans and animals, including foods produced from genetically engineered (GE) plants. Foods from GE plants must meet the same food safety requirements as foods derived from traditionally bred plants.

[...]

Crop improvement happens all the time, and genetic engineering is just one form of it. We use the term "genetic engineering" to refer to genetic modification practices that utilize modern biotechnology. In this process, scientists make targeted changes to a plant's genetic makeup to give the plant a new desirable trait. For example, two new apple varieties have been genetically engineered to resist browning associated with cuts and bruises by reducing levels of enzymes that can cause browning.

Humans have been modifying crops for thousands of years through selective breeding. Early farmers developed cross breeding methods to grow numerous corn varieties with a range of colors, sizes, and uses. For example, the garden strawberries that consumers buy today resulted from a cross between a strawberry species native to North America and a strawberry species native to South America.

Source: U.S. Food & Drug Administration

A Closer Look: Gene Therapy

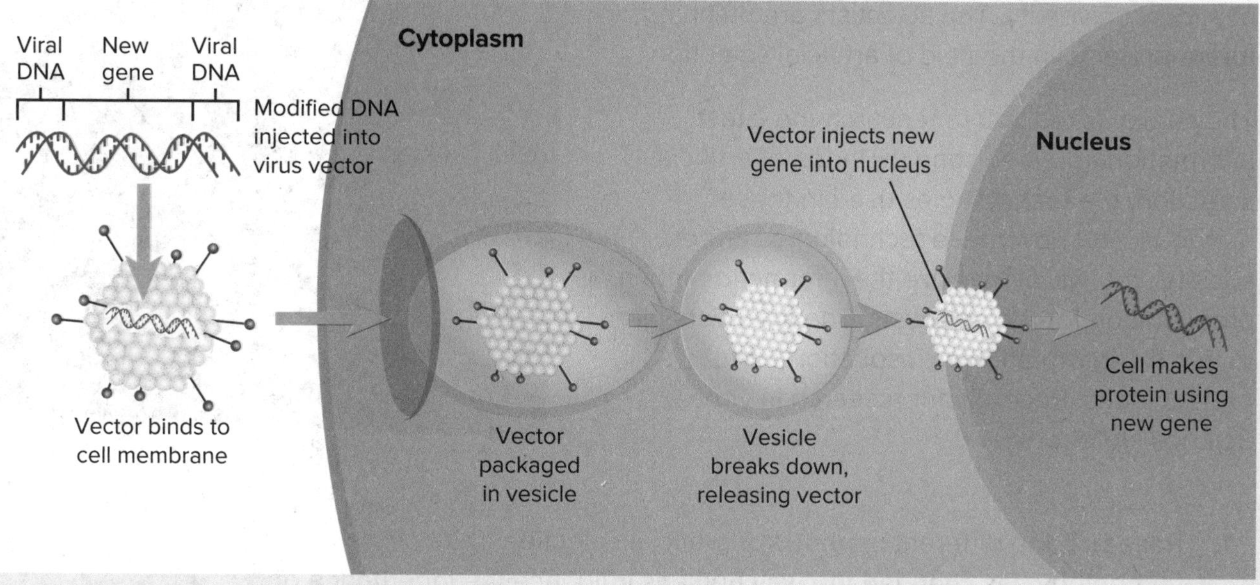

Viral DNA
New gene
Viral DNA
Modified DNA injected into virus vector
Cytoplasm
Vector binds to cell membrane
Vector packaged in vesicle
Vesicle breaks down, releasing vector
Vector injects new gene into nucleus
Nucleus
Cell makes protein using new gene

Gene therapy is the replacement of missing or malfunctioning genes by the addition of new genes to a patient's cells. DNA with the desired functioning gene can be inserted into a patient's cells instead of using medications or surgery. The modified DNA is placed in a vector to deliver it to the nucleus of a cell. The introduction of functioning genes can replace or inactivate faulty or mutated genes or provide a new gene to help fight disease.

The treatment was first tested in humans in 1990 and has continually been researched. Gene therapy has potential for treating cancer, inherited disorders, and other diseases.

In 2017, the United States Food and Drug Administration made the first gene therapy available in the U.S. The approved treatment is used for pediatric and young adult patients with a form of acute lymphoblastic leukemia, which is a cancer of the blood.

It's Your Turn

WRITING Connection With a partner, conduct additional research on gene therapy. Prepare a short segment for the school news station explaining the impacts of genetic engineering on society.

INVESTIGATION

Changing the Future

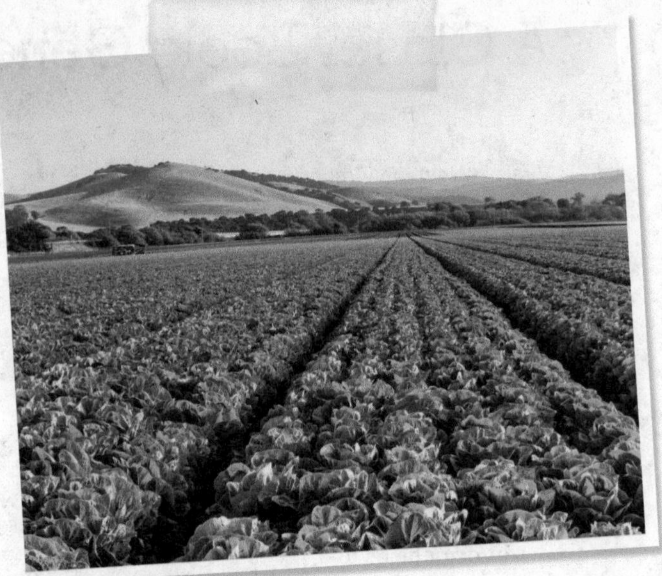

Imagine you are an intern for a group of scientists at a local university. The scientists are planning future projects in the field of artificial selection.

They have asked you to research the latest information about existing methods of artificial selection, the technologies used in these methods, and how these technologies affect society. You will synthesize the information into a presentation for the scientists. The scientists plan to use the information as inspiration for future project ideas. Record your research in your Science Notebook.

1. Research the different methods of artificial selection and how technology has changed the way humans influence the inheritance of desired traits in organisms. Choose one method of artificial selection for your research. Record and describe your choice below.

2. Develop a research plan in your Science Notebook for your chosen method. Also, recall what is required for your project—a description of the latest information about your chosen method, the technologies used in your chosen method, and how the technologies affect society.

3. Carry out your research. Be sure to use multiple sources. When you gather information from a source, be sure to record information about the credibility of the source. Identify your sources below.

4. In your Science Notebook, develop a plan to present your findings.

5. Create your presentation. Be sure your presentation shows information in a clear, logical, and engaging manner.

6. Give your presentation to your class, who will be acting as the university scientists.

7. What are the strengths and weaknesses of the project? What revisions would improve your project?

8. Will your project meet the needs of researchers? Explain your answers.

9. Will your findings allow the university scientists to completely understand current technologies and also help them plan future projects? Why or why not?

10. While you can provide the scientists with information, what else may influence their decision?

COLLECT EVIDENCE

What evidence have you discovered for how humans influence traits of organisms, such as the dogs at the beginning of the lesson, through genetic engineering? Record your evidence (B) in the chart at the beginning of the lesson.

LESSON 3

Review

Summarize It!

1. **Write** a paragraph explaining how natural selection and selective breeding are related. Include a main idea, supporting details, and a concluding sentence.

Three-Dimensional Thinking

A student prepared this chart comparing examples of natural selection with artificial selection.

Natural Selection Traits That Benefit the Species	Artificial Selection Traits That Benefit Humans
• Ability to escape predators	•
• Ability to resist droughts	•

2. Which can the student add in the column under artificial selection to complete the chart?

1. ability to grow large kernels of corn
2. ability to survive cold temperatures
3. ability to attract pollinators
4. ability to produce milk for offspring

A 1 and 3

B 1 and 2

C 2 and 3

D 3 and 4

---

3. Golden rice is a type of rice that has been altered to contain vitamin A. This yellow rice is beneficial to populations that typically do not receive enough vitamin A from other sources. How is golden rice classified?

A genetically engineered

B genetically modified organism

C altered through gene therapy

D A and B

Real-World Connection

4. **Evaluate** Imagine you and your classmates are growing a class garden. What traits would be beneficial for your plants based on the environment around your school? How would you encourage these traits?

Still have questions?
Go online to check your understanding of artificial selection.

REVISIT

Do you still agree with the person you chose at the beginning of the lesson? Return to the Science Probe at the beginning of the lesson. Explain why you agree or disagree with that person now.

EXPLAIN THE PHENOMENON

Revisit your claim about how humans can control traits of organisms. Review the evidence you collected. Explain how your evidence supports your claim.

PLAN AND PRESENT

STEM Module Project
Science Challenge

Now that you've learned artificial selection, go back to your Module Project to finish planning and conduct your activity. You'll want to think about how artificial selection is similar to natural selection in organisms, for example, the owl butterfly at the beginning of the module.

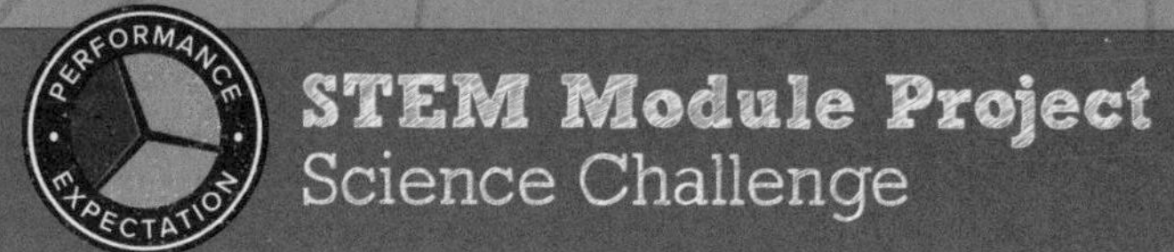

Population Probabilities

The education director of an environmental studies lab has asked you to help her design and test an activity. The activity will combine math and science to explain how natural selection changes the frequency of a trait in a population over time.

Your task is to design and perform the activity and then to construct an explanation of the resulting data. The education director has explained that the activity will involve deer mice, which can be either brown or gray.

Planning After Lesson 1

What can you infer about the genetics of the deer mice you will use in your model, based on the observation that they have two different physical traits for fur color?

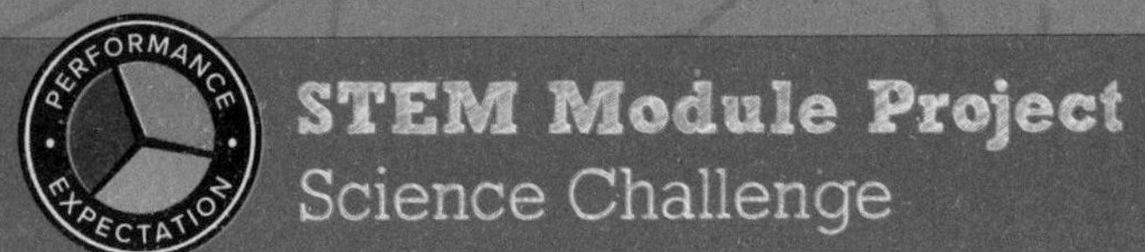

Planning After Lesson 1, continued

In the space below, make a model, a labeled diagram, or a flow chart, that describes the relationship between chromosomes, genes, proteins and their structures, and the observable trait of fur color in deer mice.

Using your model above as a basis for your explanation, describe how a mutation could cause a change in the fur color of a deer mouse. Then, explain how a mutation that impacts fur color might be included in your activity.

Planning After Lesson 2

The mathematical activity you develop will include two variations in fur color of deer mice, gray fur and brown fur. Explain how fur color could be an adaptation in deer mice.

Identify several environmental factors that could affect which mice in a population survive to reproduce. These will be factors that can be included in your activity.

Predict how the proportion of a population with an advantageous trait will change over several generations in your activity. Explain your prediction.

Predict what would happen if your activity included an environmental shift that is so extreme or rapid that the population does not have time to adapt.

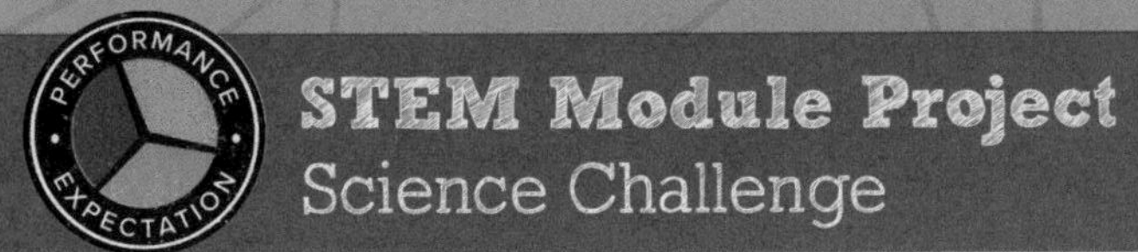

Planning After Lesson 3

In your activity, you will model natural selection. Describe one way in which natural selection and selective breeding are similar.

Could your activity be modified to model artificial selection? Explain.

Could your activity model what might happen if genetically modified deer mice with white fur were accidentally released into the wild population of mice? Explain.

Complete a Sample Activity

The educational director has provided the following sample activity as one possible way to model the effect of natural selection of a population. Carry out this activity and gather data.

This activity will model changes in the distribution of fur colors within a population of deer mice. Carry out a series of 50 coin flips (assign one color as "heads" and the other as "tails") to describe the distribution of fur-color trait in the original population of deer mice.

Brown: ______________________ Gray: ______________________

Read the following scenario:

A drought kills much of the grass in the field where your original population of mice live, leaving large patches of bare brown soil. Brown mice have an advantageous trait in this environment, and have a higher probability of surviving and reproducing. For each brown mouse in this population, an average of three brown offspring are produced per generation. Gray mice have a lower probability of surviving and reproducing. For each gray mouse in this population, two gray offspring are produced per generation. Mice have a short life span so, for each generation, count only the newly produced offspring.

Using the data about your original population and the information in the scenario, complete the following:

	Brown Mice	**Gray Mice**
Original population		
Generation 2		
Generation 3		
Generation 4		
Generation 5		

Examine the data. How did the characteristics of the population change over time in response to environmental change? Record your answer in your Science Notebook.

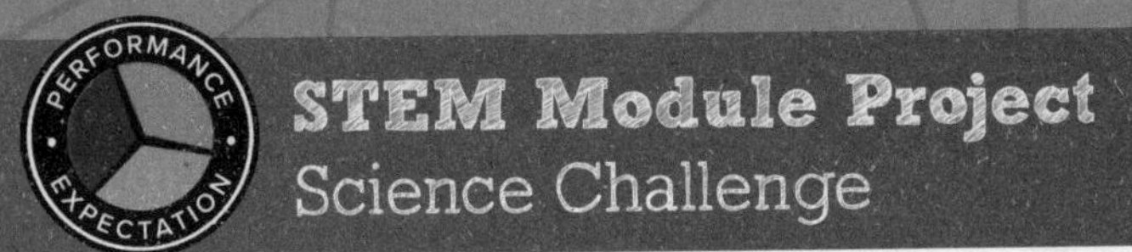

Develop Your Activity

In your Science Notebook, develop your own mathematical activity about brown and gray deer mice. Be sure to identify the following components:

- specific environmental factor or change impacting the population
- which trait is advantageous and why, and how that affects probability of survival and reproduction, and
- how many offspring are produced by each type of mouse each generation.

Carry out the activity as described in your scenario. Record your data in the space below.

	Brown Mice	**Gray Mice**
Original population		
Generation 2		
Generation 3		
Generation 4		
Generation 5		

Did your data provide evidence that populations adapt through natural selection? Explain your answer in your Science Notebook.

Construct Your Explanation

Develop a plan for the explanation you will provide to the education director. Think about the following questions to help you decide what kind of information to include and how to present it. Place a check mark next to each question after your group has discussed it.

- ______ How did you determine the number of mice of each color in your original population?

- ______ How many mice of each color were in the starting and final populations? What were the percentages? What were the ratios?

- ______ If you were able to carry out the activity more than once, were your results consistent? If not, why?

- ______ What trends did you notice in the distribution of traits in this population?

- ______ What cause-and-effect relationships between environmental conditions and natural selection did you identify?

- ______ Can you use the data from your mathematical activity to support an explanation that natural selection may lead to increases and decreases of specific traits in populations over time?

- ______ Could your mathematical activity model what would happen if an environmental change occurred too quickly for a population to adapt?

- ______ What are the limitations of your activity? For example, is your activity designed to predict with 100 percent certainty what would happen in a natural environment with multiple cause-and-effect relationships affecting a population?

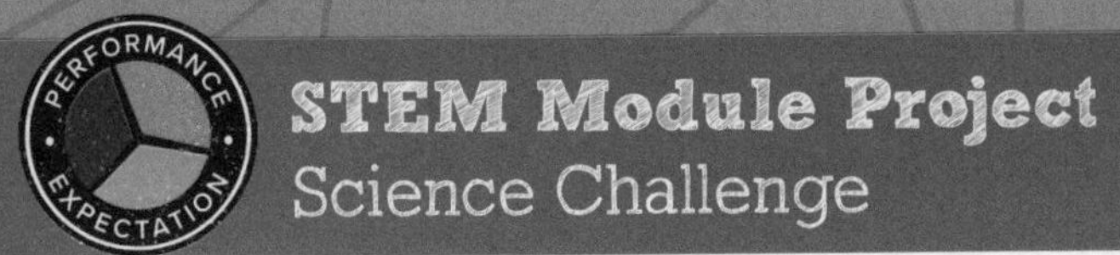

Create Your Presentation

Complete and finalize your presentation.

Be sure to include an explanation based on evidence from your mathematical activity that describes:

- How genetic variations of traits in a population affect an individual organism's probability of surviving and reproducing in an environment.
- How natural selection may lead to increases and decreases of specific traits in populations over time.

Present your report.

How did your mathematical activity compare to what you have read about in this module?

Identify one strength and one weakness of your activity or report.

How does your research into natural selection and traits help you understand why some butterflies have coloration that looks like the face of a hungry owl?

Congratulations! You've completed the Science Challenge requirements!

Module Wrap-Up

REVISIT
THE PHENOMENON

Using the concepts that you have learned throughout this module, explain why the butterfly looks like the face of an owl.

INQUIRY

If you had to ask one question about what you studied, what would it be?

Plan and conduct an investigation to answer this question.

Evidence of Evolution

ENCOUNTER THE PHENOMENON

How do whale pelvic bones show evidence of evolution and what other evidence for evolution exists?

Whale of an Evolutionary Tale

GO ONLINE
Watch the video *Whale of an Evolutionary Tale* to see this phenomenon in action.

Collaborate With a partner, discuss pelvic bones. Why do you think the whale has pelvic bones even though it does not have legs? Record or illustrate your thoughts in the space below.

STEM Module Project Launch
Science Challenge

It's All Relative

The director of a natural history museum wants to develop a series of exhibits that show how present-day organisms are related to ancestral fossil organisms and to other present-day organisms.

He asks you to prepare an exhibit about one specific modern organism and its relationships to fossilized organisms and modern organisms. Along with your exhibit, the director has asked you to prepare a written explanation for museum visitors. You will need to explain how the fossil and biological evidence can be used to make inferences about evolutionary relationships.

Lesson 1
Fossil Evidence of Evolution

Lesson 2
Biological Evidence of Evolution

Start Thinking About It

What scientific ideas might you include in an explanation about similarities and differences between extinct and modern organisms?

STEM Module Project

Planning and Completing the Science Challenge How will you meet this goal? The concepts you will learn throughout this module will help you plan and complete the Science Challenge. Just follow the prompts at the end of each lesson!

LESSON 1 LAUNCH

SCIENCE PROBES

Endless Fossi-bilities

Four friends were comparing their ideas about the meaning of fossils. This is what they said:

Emma: I think fossils are pieces of dead animals and plants, and tell us little about the animal or plant.

Aidan: I think fossils only come from bones of extinct animals that lived millions of years ago.

Ethan: I think fossils are the evidence of the existence of animals which are seen in the remains of bones, shells, or even impressions of rock layers.

Madison: Fossils are the remains of plants and animals that have recently died. Their remains cannot be preserved for very long.

With whom do you agree most? Explain why you agree with that person.

You will revisit your response to the Science Probe at the end of the lesson.

LESSON 1

Fossil Evidence of Evolution

ENCOUNTER THE PHENOMENON | How do fossils, such as *Tiktaalik*, provide evidence of evolution?

Observe the fossil in the photo. Draw what you think the organism looked like while it was living.

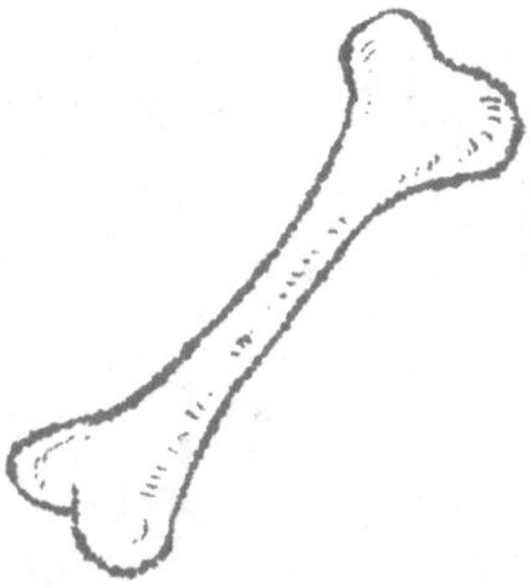

GO ONLINE
Check *Terrific Tiktaalik* to see this phenomenon in action.

Terrific Tiktaalik

EXPLAIN THE PHENOMENON

Did you notice that the *Tiktaalik* fossil looked familiar, but still very different from animals that you see every day? Use your observations about the phenomenon to make a claim about what the *Tiktaalik* fossil can tell you about present-day organisms and how they evolved.

CLAIM

The *Tiktaalik* fossil offers evidence of evolution...

COLLECT EVIDENCE

as you work through the lesson. Then return to these pages to record your evidence.

EVIDENCE

A. What evidence have you discovered that explains what fossils, like the *Tiktaalik*, can tell us about time?

MORE EVIDENCE

B. What evidence have you discovered that explains how patterns in the fossil record provide evidence of evolution?

When you are finished with the lesson, review your evidence. If necessary, based on the evidence, revise your claim.

REVISED CLAIM

The *Tiktaalik* fossil offers evidence of evolution...

Finally, explain your reasoning for how and why your evidence supports your claim.

REASONING

The evidence I collected supports my claim because...

How do fossils form?

You might already know that fossils are the remains or evidence of once-living organisms. Evidence from fossils helps scientists understand how organisms have changed over time. When scientists find fossils they are often underground or within rock layers. How did they get there and how did they form?

When you think about fossils, you might picture the large skeletons of dinosaurs in museums. In fact, there are different types of fossils that form under different circumstances. Let's investigate one of those ways.

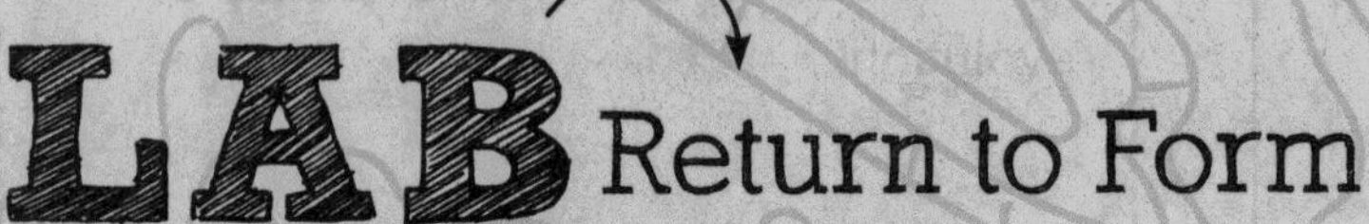

Safety 

Materials

- container of moist sand
- shell
- newspaper
- plaster of paris

Procedure

1. Read and complete a lab safety form.
2. Place a container of moist sand on top of newspaper. Press a shell into the moist sand. Carefully remove the shell. Brush any sand on the shell into the newspaper.
3. Pour plaster of paris into the impression. Wait for it to harden.
4. Remove the shell fossil from the sand and brush it off.
5. Observe the structure of the fossil.
6. Follow your teacher's instructions for proper cleanup.

Analyze and Conclude

7. What effect did the shell have on the sand?

Fossil Formation When plants and animals die, any soft tissues animals do not eat are broken down by bacteria. Only a dead animal's hard parts, such as bones, shells, and teeth, remain. In most instances, these hard parts also break down over time. However, under rare conditions, some become fossils. The soft tissues of animals and plants, such as skin, muscles, or leaves, can also become fossils, but these are even rarer. Some fossils form when impressions left by organisms in sand or mud are filled in by sediments that harden. You observed this in the Lab *Return to Form*. Examine the table below to see some ways that fossils can form.

Want more information?
Go online to read more about fossil evidence of evolution.

Fossil Type	Description	Example
Mineralization	Rock-forming minerals, such as calcium carbonate ($CaCO_3$), in water filled in the small spaces in the tissue of these pieces of petrified wood. Water also replaced some of the wood's tissue. Mineralization can preserve the internal structures of an organism.	
Carbonization	In carbonization, pressure drives off a dead organism's liquids and gases. Only the carbon outline, or film, of the organism remains. Fossil films made by carbonization are usually black or dark brown. Fish, insects, and plant leaves, such as these fern fronds, are often preserved as carbon films.	
Molds and Casts	When sediments hardened around this buried trilobite, a mold formed. Molds usually show hard parts, such as shells or bone. If a mold is later filled with more sediments that harden, the mold can form a cast.	
Trace Fossils	A trace fossil is the preserved evidence of the activity of an organism. These footprints were made when a dinosaur walked across mud that later hardened. This trace fossil might provide evidence of the speed and weight of the dinosaur.	
Original Material	If original tissues of organisms are buried in the absence of oxygen for long periods of time, they can fossilize. The insect in this amber became stuck in tree sap that later hardened.	

What can fossils tell us about time?

On your way to school, you might have seen an oak tree or heard a robin. Although these organisms shed leaves or feathers, their characteristics remain the same from day to day. It might seem as if they have been on Earth forever. However, if you were to travel back in time, you might not see oak trees or robins. You may see different species of trees and birds. That is because species change over time.

FOLDABLES®

Go to the Foldables® library to make a Foldable® that will help you take notes while reading this lesson.

INVESTIGATION

Guess the Age

In the image you can see fossils buried in rock layers. Examine the image and answer the questions below.

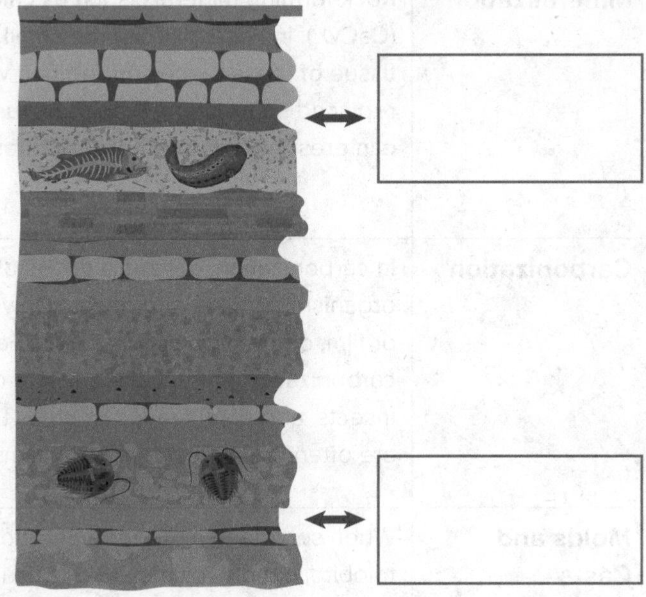

1. If the topmost rock layer of the image is present day, then what is the relative age of the areas that are indicated by the arrows to each other? Infer the age of the areas by writing *older* or *younger* in the boxes provided.

2. Why did you place the words *older* or *younger* in those locations?

3. What do you think fossils can tell us about time?

The Fossil Record You may have guessed that the fossils at the bottom of the image were much older than the fossils at the top. The fossils at the bottom were left by animals that existed long before the fossils at the top.

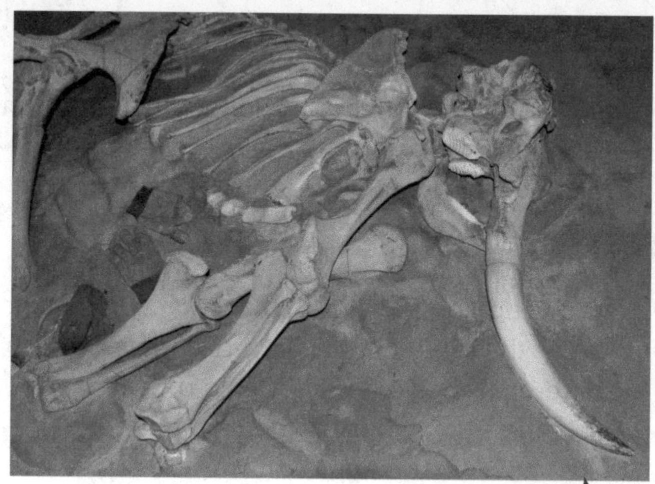

Pygmy mammoth fossils have been found only in California.

These fossils help make up the fossil record. The **fossil record** is made up of all the fossils ever discovered on Earth. It contains millions of fossils that represent thousands of species. Most of these species are no longer alive on Earth. The fossil record provides evidence that species have changed over time. Using the fossil record, scientists are able to determine when an organism lived.

Scientists cannot date most fossils directly. Instead, they date the rocks the fossils are embedded inside. Rocks erode or are recycled over time. However, scientists can determine ages for most of Earth's rocks. Examine the table below to learn how scientists determine the age of fossils.

Relative-Age Dating	Absolute-Age Dating
• In relative-age dating, scientists determine the relative order in which rock layers were deposited. In an undisturbed rock formation, they know that the bottom layers are oldest and the top layers are youngest. • Relative-age dating helps scientists determine the relative order in which species have appeared on Earth over time.	• Absolute-age dating is more precise than relative-age dating. Scientists take advantage of radioactive decay, a natural clock-like process in rocks, to learn a rock's absolute age, or its age in years. • In radioactive decay, unstable isotopes in rocks change into stable isotopes over time. Scientists measure the ratio of unstable isotopes to stable isotopes to find the age of a rock.

Geologic Time It is hard to keep track of time that is billions of years long. Evidence of microscopic, unicellular organisms has been found in rocks 3.4 billion years old. Scientists organize Earth's history into a time line called the geologic time scale. The **geologic time scale** is a chart that divides Earth's history into different time units. The longest time units in the geological time scale are eons. Earth's history is divided into four eons. Earth's most recent eon—the Phanerozoic (fa nuh ruh ZOH ihk) eon—is subdivided into three eras. Examine the geologic time scale on the next page.

The Geologic Time Scale Look at the figure of the geologic time scale. You might have noticed that neither eons nor eras are equal in length. When scientists began developing the geologic time scale in the 1800s, they did not have absolute-age dating methods. To mark time boundaries, they used fossils. Fossils provided an easy way to mark time. Scientists knew that different rock layers contained different types of fossils. Some of the fossils scientists used to mark the time boundaries are shown in the figure.

Often, a type of fossil found in one rock layer did not appear in layers above it. Even more surprising, entire collections of fossils in one layer were sometimes absent from layers above them. It seemed as if whole communities of organisms had suddenly disappeared.

The Geologic Time Scale

Millions of years ago	Eons	Eras	Periods
0–2.6	Phanerozoic	Cenozoic	Quaternary
2.6–23	Phanerozoic	Cenozoic	Neogene
23–66	Phanerozoic	Cenozoic	Paleogene
66–145	Phanerozoic	Mesozoic	Cretaceous
145–201	Phanerozoic	Mesozoic	Jurassic
201–252	Phanerozoic	Mesozoic	Triassic
252–299	Phanerozoic	Paleozoic	Permian
299–359	Phanerozoic	Paleozoic	Carboniferous
359–419	Phanerozoic	Paleozoic	Devonian
419–444	Phanerozoic	Paleozoic	Silurian
444–485	Phanerozoic	Paleozoic	Ordovician
485–541	Phanerozoic	Paleozoic	Cambrian
541–2,500	Precambrian	Proterozoic	
2,500–4,000	Precambrian	Archean	
4,000–4,600	Precambrian	Hadean	

Extinctions Scientists now understand that sudden disappearances of fossils in rock layers are evidence of extinction events. **Extinction** (ihk STINGK shun) occurs when the last individual organism of a species dies. A mass extinction occurs when many species become extinct within a few million years or less.

The fossil record contains evidence that five mass extinction events have occurred during the Phanerozoic eon. Extinctions also occur at other times, on smaller scales. Evidence from the fossil record suggests extinctions have been common throughout Earth's history. Examine the timeline of extinction events on the next page.

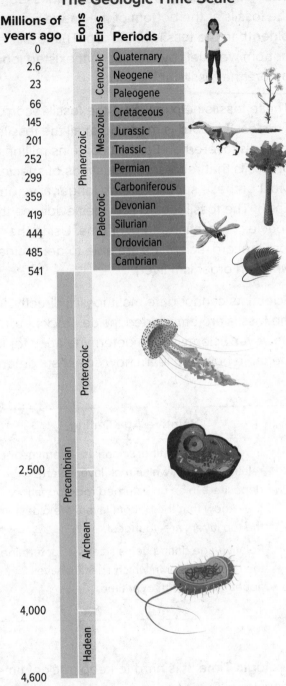

GO ONLINE for additional opportunities to explore!

EARTH SCIENCE Connection Investigate what fossils can tell us about time by performing one of the following activities.

☐ To learn more about Earth's history, watch the **Personal Tutor** *Geologic Time Scale.*

OR

☐ To learn more about how fossils are dated, explore the **Virtual Lab** *How can fossil and rock data determine when an organism lived?*

Extinction Events

Late Ordovician
Late Devonian
Late Permian
Late Triassic
Late Cretaceous

Number of Genera (singular Genus): 4,000; 3,000; 2,000; 1,000

Millions of Years Ago (mya): 500; 400; 300; 200; 100; 0

This graph shows the five major extinctions of the Phanerozoic era!

Environmental Change What causes extinctions? Populations of organisms depend on resources in their environment for food and shelter. Sometimes environments change. After a change happens, individual organisms of a species might not be able to find the resources they need to survive. When this happens, the organisms die, and the species becomes extinct. In the figure above, you can examine major extinction events that have occurred over time. Learn about the two types of environmental change below.

Sudden Changes	Extinctions can occur when environments change quickly. A volcanic eruption or a meteorite impact can throw ash and dust into the atmosphere, blocking sunlight for many years. This can affect global climate and food webs. Scientists hypothesize that the impact of a huge meteorite 65 million years ago contributed to the extinction of dinosaurs.
Gradual Changes	Not all environmental change is sudden. Depending on the location, Earth's tectonic plates move between 1 and 15 cm each year. As plates move and collide with each other over time, mountains form and oceans develop. If a mountain range or an ocean isolates a species, the species might become extinct if it cannot find the resources it needs. Species also might become extinct if sea level changes.

THREE-DIMENSIONAL THINKING

With a partner, think of the **effects** of environmental change that may have **caused** extinction in your local ecosystem. Write a script for a podcast in which you discuss and explain the extinction. Use your Science Notebook if extra writing space is required.

COLLECT EVIDENCE

What can fossils, like the *Tiktaalik*, tell us about time? Record your evidence (A) in the chart at the beginning of the lesson.

How do fossils show change over time?

The fossil record contains clear evidence of the extinction of species over time. But it also contains evidence of the appearance of many new species. How do different species arise?

LAB It's Time for a Change

Over long time periods on Earth, certain individuals within populations of organisms were able to survive better than others.

Safety

Materials

species I.D. cards
chart paper
colored pencils
markers

Procedure

1. Choose a species from the Species I.D. Cards.
2. **ART Connection** On chart paper, draw six squares in a row and number them 1–6 respectively. Use colored pencils and markers to make a comic strip showing the ancestral and present-day forms of your species in frames 1 and 6.
3. Use information from the I.D. Card to show what you think would be the progression of changes in the species in frames 2–5.
4. In speech bubbles, explain how each change helped the species survive.

Analyze and Conclude

5. Infer why a scientist would identify a fossil from the species in the first frame of your cartoon as the ancestral form of the present-day species.

6. How would the fossils of the species at each stage provide evidence of biological change over time?

Extinctions and Evolution Many early scientists thought that each species appeared on Earth independently of every other species. However, as more fossils were discovered, patterns in the fossil record began to emerge. Many fossil species in nearby rock layers had similar body plans and similar structures. It appeared as if they were related. For example, the series of horses in the figure suggests that the modern horse is related to other extinct species. These species changed over time in what appeared to be a sequence. Recall that change over time is evolution. Biological evolution is the change over time in populations of related organisms.

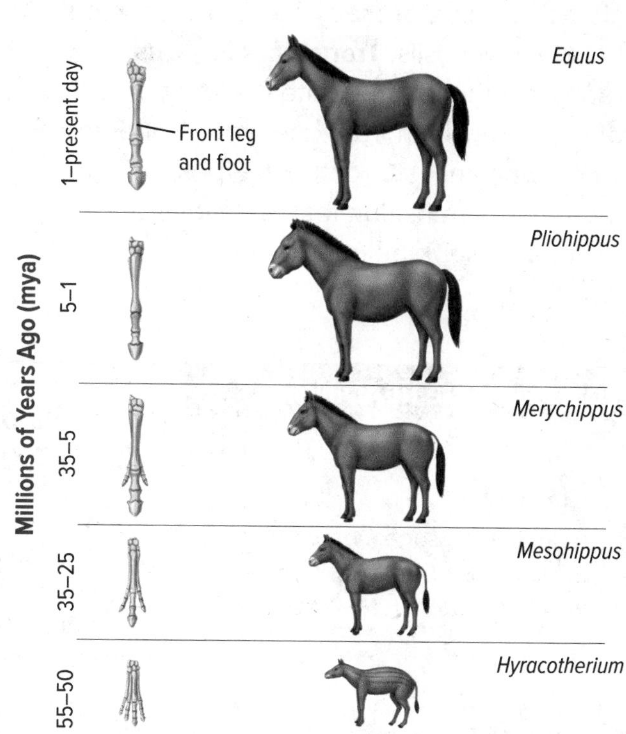

THREE-DIMENSIONAL THINKING

Examine the chart of the evolution of horses above. What has changed over 55 million years and what do you think drove that **change?**

COLLECT EVIDENCE

What can patterns in the fossil record tell us about evolution? Record your evidence (B) in the chart at the beginning of the lesson.

Read a Scientific Text

In the chart on the previous page, you saw a significant change in horses over 55 million years. The smaller *Hyracotherium* with short legs became the large *Equus* with long legs we know today. This change occurred because more modern horses needed to be able to reach greater speeds to outrun predators.

In the chart, you also may have noticed a pattern as the horse seemed to gradually change as time passed. These represented transitional fossils. **Transitional fossils** represent the intermediate evolutionary forms of life in transition from one type to another, or a common ancestor of these types. To learn more about transition fossils, read about *Tiktaalik roseae* below.

Tiktaalik roseae was a transitional species that lived 375 million years ago.

CLOSE READING

Inspect

Read the passage *Details of Evolutionary Transition from Fish to Land Animals Revealed.*

Find Evidence

Reread the passage. Underline phrases that support evidence of *Tiktaalik roseae* being a transitional fossil.

Make Connections

Communicate With your partner, research the evolutionary history of *Tiktaalik rosea*. Drawing evidence from the text to support your analysis, reflection, and research, create a slideshow based on your findings.

Details of Evolutionary Transition from Fish to Land Animals Revealed

New research has provided the first detailed look at the internal head skeleton of *Tiktaalik roseae*, the 375-million-year-old fossil animal that represents an important intermediate step in the evolutionary transition from fish to animals that walked on land.

[...]

The body plan and nature of the deposits where the fossils were found suggest an animal that lived on the bottom in shallow water, and perhaps out of the water for short periods.

Tiktaalik roseae has features of the skull, neck, ribs and appendages that are shared with the earliest limbed animals (tetrapods), as well as fishlike features such as scales and fin rays. This mosaic of features makes it a textbook example of a transitional fossil, say paleontologists.

[...]

"The braincase, palate and gill arches of Tiktaalik help reveal the pattern of evolutionary change in this part of the skeleton," said Downs. "We see that cranial features once associated with land-living animals were in fact the first adaptations for life in shallow water."

Source: National Science Foundation

STEM CAREER Connection

A Day in the Life of a Paleontologist

These paleontologists are digging up a fossilized mammoth skeleton from the Ice Age!

EARTH SCIENCE Connection A paleontologist is a scientist who studies fossils. They do this in order to learn about what they can teach us about life on Earth in a different time. You may think of all paleontologists as just studying dinosaur bones, but actually, there are many different kinds of paleontologists. Some study fish or mammals. Some even study plants!

Paleontologists may work with research organizations like universities. They may also work in government or with private businesses. Paleontologists could work in an office or lab. Often times, they are busy uncovering fossils in the field. They travel to different locations to discover and study fossils. Their work may be physically demanding at times. Living conditions can be harsh and the weather may range from intense heat to extreme cold. Paleontologists may also be found working in a laboratory, studying their discoveries.

It's Your Turn

WRITING Connection Suppose you are a paleontologist working with the developer of a planned community. Make a slideshow presentation for the developer explaining the impact that human development can have on potential fossil discoveries, as well as the importance of such discoveries.

LESSON 1

Review

Summarize It!

1. Based on what you have read, create a graphic organizer detailing what the fossil record can tell us about present-day organisms.

Three-Dimensional Thinking

Use the figure below to answer question 2.

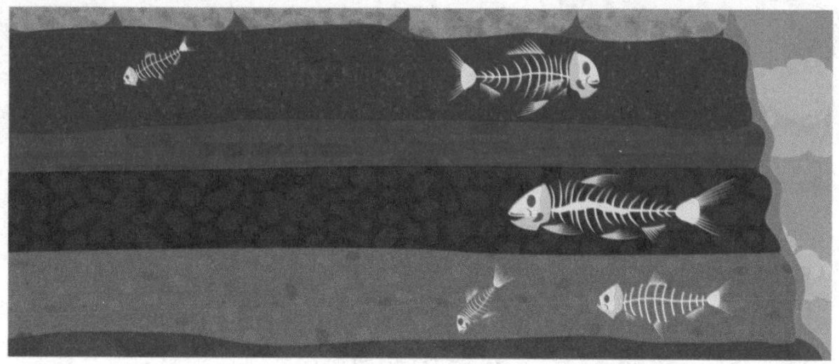

2. What can scientists use to explain when the fossils in the bottom of the figure appeared on Earth?

   A relative-age dating

   B trace fossils

   C mineralization

   D rock layers

---

3. Which is an example of a sudden change that could explain the extinction of a species?

   A a mountain range isolates a spices

   B Earth's tectonic plates move

   C a volcano erupts

   D sea level changes

Real-World Connection

4. **Infer** Suppose you found a fossil that not only had fins and gills, but also lungs and wrists. What might this fossil suggest about evolution?

5. **Explain** Your fellow classmate says that fossils are just pieces of dead animals. Explain why you agree or disagree with them.

Still have questions?
Go online to check your understanding about fossil evidence of evolution.

REVISIT SCIENCE PROBES

Do you still agree with the person you chose at the beginning of the lesson? Return to the Science Probe at the beginning of the lesson. Explain why you agree or disagree with that person now.

EXPLAIN THE PHENOMENON

Revisit your claim about what fossils can tell us about present-day organisms and how they evolved. Review the evidence you collected. Explain how your evidence supports your claim.

START PLANNING

STEM Module Project Science Challenge

Now that you've learned about fossil evidence of evolution, go back to your Module Project to plan your exhibit. You'll want to explain how the pelvic bones in whales show a relationship between fossils of organisms that lived long ago and present-day organisms.

PERFORMANCE EXPECTATION

SCIENCE PROBES

LESSON 2 LAUNCH

Fins and Flippers

Three friends were comparing the fin of a fish with the flipper of a whale. This is what they said:

Emily: I think that the fin of the fish and the flipper of the whale have the same structure, but different functions.

Ariana: I think that the fin of the fish and the flipper of the whale have a similar function, but differ in structure.

Jayden: I think that the fin of the fish and the flipper of the whale are too different to even compare.

Circle the name of the person you agree with most. Explain why you agree with that person.

You will revisit your response to the Science Probe at the end of the lesson.

LESSON 2

Biological Evidence of Evolution

ENCOUNTER THE PHENOMENON

How can the wings of a bat and the wings of a bird provide evidence for evolutionary relationships?

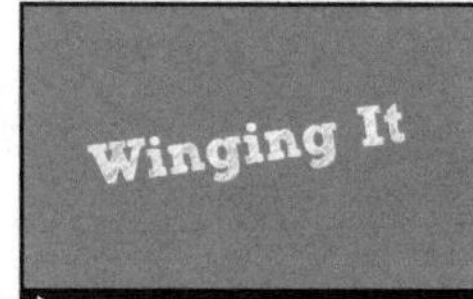

GO ONLINE Check out *Winging It* to see the phenomenon in action.

Observe the two animals in the video and sketch them in the space below. Compare structures using labels on your sketches.

What might the similarities or differences between the two animals tell you about their evolutionary history?

EXPLAIN THE PHENOMENON

You just observed two organisms in flight—a bat and a bird. Both animals have wings but they are very different from one another. Make a claim about their evolutionary relationship based on these differences and similarities.

CLAIM

The evolutionary relationship of bats and birds...

COLLECT EVIDENCE as you work through the lesson. Then return to these pages to record your evidence.

EVIDENCE

A. What evidence have you discovered that explains how comparative anatomy helps with understanding the evolutionary history of bats and birds?

MORE EVIDENCE

B. What evidence have you discovered that explains how embryology helps with understanding the evolutionary history of bats and birds?

C. What evidence have you discovered that explains how molecular biology helps with your understanding of the evolutionary history of bats and birds?

When you are finished with the lesson, review your evidence. If necessary, based on the evidence, revise your claim.

REVISED CLAIM

The evolutionary relationship of bats and birds...

Finally, explain your reasoning for how and why your evidence supports your claim.

REASONING

The evidence I collected supports my claim because...

How are a human's arm and a bird's wing related?

Often times, the way something is structured, like the wing of a bird or bat, can help give insight into its function. Explore some common objects to compare structure and function in the following lab.

Want more information?
Go online to read more about biological evidence of evolution.

Go to the Foldables® library to make a Foldable® that will help you take notes while reading this lesson.

Spoon Something Up

Safety 

Materials

set of spoons

Procedure

1. Read and complete a lab safety form.
2. In a small group, examine your set of spoons and discuss your observations.
3. Sketch or describe the structure of each spoon in your Science Notebook. Discuss the purpose that each spoon shape might serve.
4. Label the spoons in your Science Notebook with their purposes.
5. Follow your teacher's instructions for proper cleanup.

Analyze and Conclude

6. If spoons were organisms, what do you think the ancestral spoon would look like? Record your responses in your Science Notebook.
7. Explain how three of the spoons have different structures and functions, even though they are related by their similarities. Record your responses in your Science Notebook.

Comparative Anatomy As you just observed, spoons that look different from one another still have some structural and functional similarities. Like spoons, animals also have similarities, even when they appear very different. Observations of structural and functional similarities and differences in species that do not look alike are possible through comparative anatomy. **Comparative anatomy** is the study of similarities and differences among structures of living species.

Homologous Structures Humans and birds look different and move in different ways. Humans use their arms for balance and their hands to grasp objects. Birds use their forelimbs as wings for flying. However, the forelimb bones of these species exhibit similar patterns. **Homologous** (huh MAH luh gus) **structures** are body parts of organisms that are similar in structure and position but different in function.

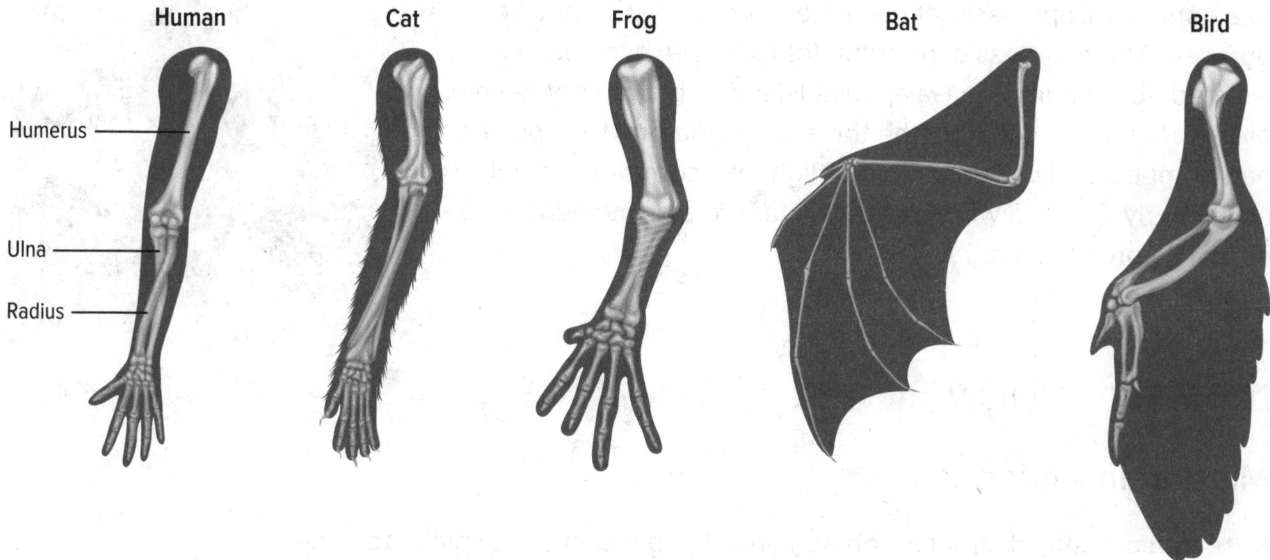

As shown in the figure, homologous structures, such as the forelimbs of humans, cats, frogs, bats, and birds, suggest that these species are related. The more similar two structures are to each other, the more likely it is that the species have evolved from a recent common ancestor.

THREE-DIMENSIONAL THINKING

Think about **patterns** in **structures**, other than the arm, that can be found across different groups of animals. In your Science Notebook, quickly sketch your structures like in the figure above.

COLLECT EVIDENCE

How does comparative anatomy help you understand the evolutionary relationship between bats and birds? Record your evidence (A) in the chart at the beginning of the lesson.

Analogous Structures Both wings shown to the right are used for flight. But bird wings are covered with feathers. Fly wings are covered with tiny hairs. Body parts that perform a similar function but differ in structure are **analogous** (uh NAH luh gus) **structures.** Analogous structures evolved separately from one another and do not share a closely related common ancestor. Differences in the structure of bird and fly wings indicate that birds and flies are not closely related.

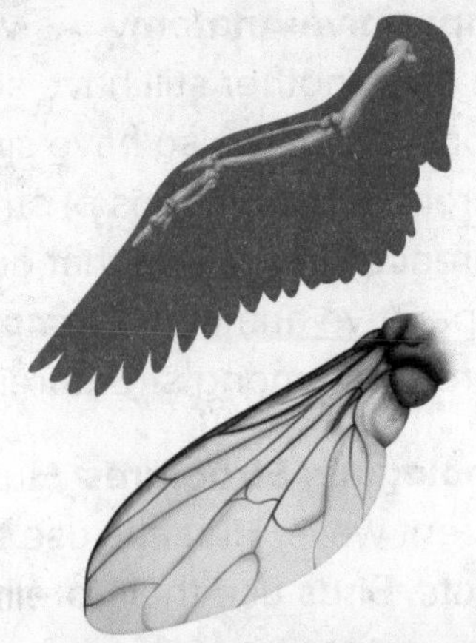

Vestigial Structures The flightless cormorant in the photo has short, stubby wings. Yet, as its name suggests, it cannot fly. The bird's wings are an example of vestigial structures. **Vestigial** (veh STIH jee ul) **structures** are body parts that have lost their original function through evolution. The best explanation for vestigial structures is that the species with a vestigial structure is related to an ancestral species that used the structure for a specific purpose. The flightless cormorant, for example, evolved in an isolated ecosystem–the Galápagos Islands–that did not have any predators on land. Without the threat of predators, the species had no need to fly. Eventually, the flightless cormorant lost its ability to fly due to a wing size that is far too small to support a bird of its proportions.

INVESTIGATION

Missing in Action

Research a vestigial structure present in a living organism. Explain how the vestigial structure arose while detailing how its function relates to its structure.

What other ways can you determine evolutionary relationships?

You just read that studying the external structures of organisms can help scientists learn more about how organisms are related. What other structures might help scientists understand how organisms are related?

INVESTIGATION

Developing Evolutionary Understanding

1. In the figure below, use different colored pencils to circle structures that are similar across all images.

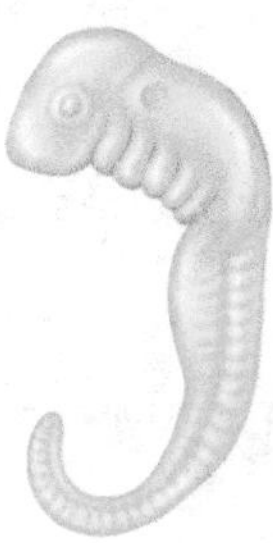

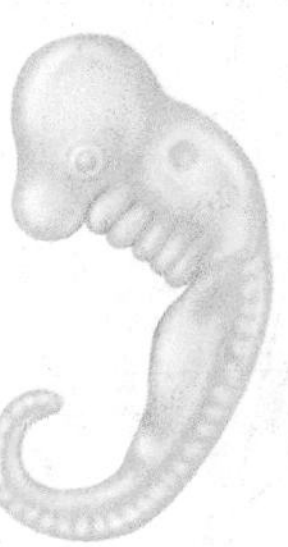

2. Describe the structures you circled and explain why you chose them.

3. Explain how these similarities may provide evidence of relatedness between the individual images.

Developmental Biology You have learned that studying the structures of organisms can help scientists learn more about how organisms are related. Studying the development of embryos can also provide scientists with evidence that certain species are related. The science of the development of embryos from fertilization to birth is called **embryology** (em bree AH luh jee).

Pharyngeal Pouches Embryos of different species often resemble each other at different stages of their development. For example, all vertebrate embryos have pharyngeal (fuh rihn JEE ul) pouches at one stage, as shown in the figure. This feature develops into different body parts in each vertebrate. Yet, in all vertebrates, each part is in the face or neck. For example, in reptiles, birds, and humans, part of the pharyngeal pouch develops into a gland in the neck that regulates calcium. In fish, the same part becomes the gills. One function of gills is to regulate calcium. The similarities in function and location of gills and glands suggest a strong evolutionary relationship between fish and other vertebrates.

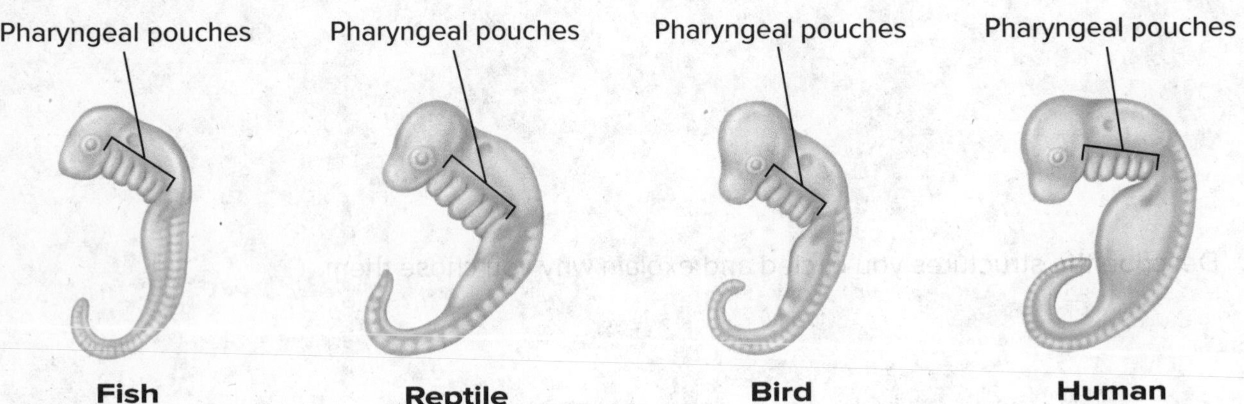

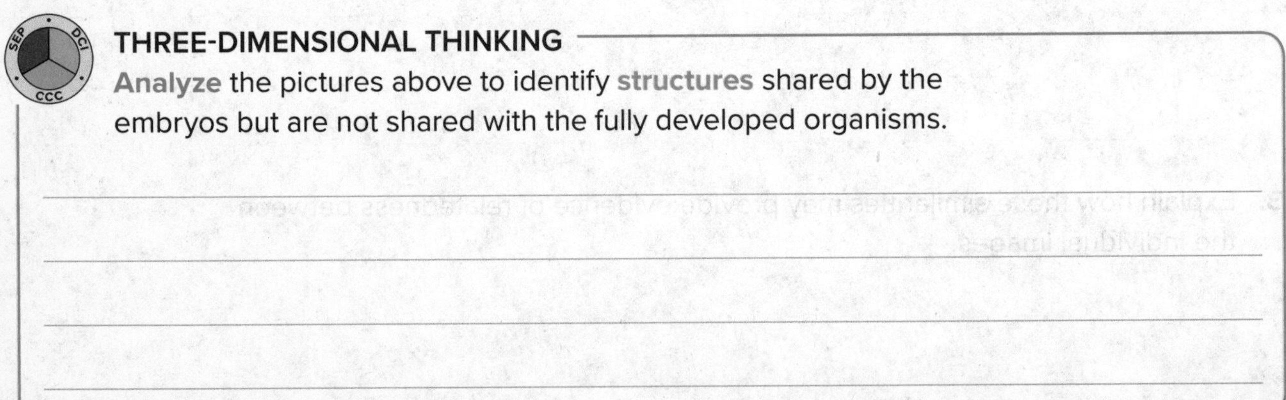

THREE-DIMENSIONAL THINKING

Analyze the pictures above to identify **structures** shared by the embryos but are not shared with the fully developed organisms.

COLLECT EVIDENCE

How does embryology help you understand the evolutionary relationship between bats and birds? Record your evidence (B) in the chart at the beginning of the lesson.

Modern Evolutionary Advances Studies of fossils, comparative anatomy, and embryology provide support for Darwin's theory of evolution by natural selection. How do you think more recent advances have allowed for greater understanding of evolution?

INVESTIGATION

Evolving Your Knowledge

Proteins, such as cytochrome *c*, are made from combinations of 20 amino acids. The graph to the right shows the number of amino acid differences in cytochrome *c* between humans and other organisms. Use the graph to answer the questions.

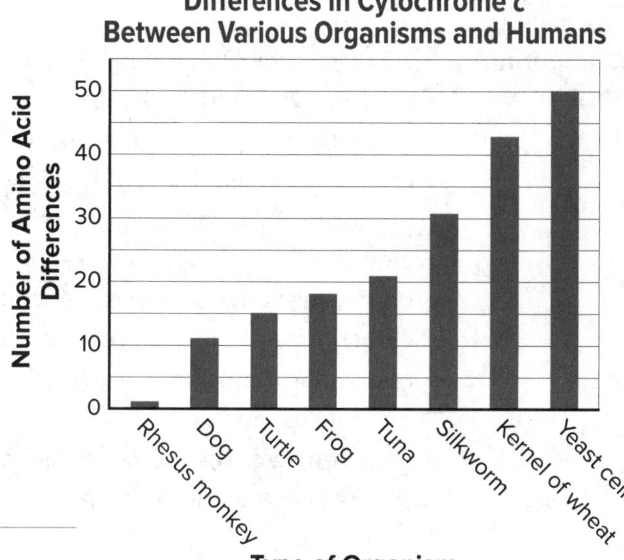

1. Which organisms do you think might be more closely related to each other: a dog and a turtle or a dog and a silkworm? Explain your answer.

2. Which organism has the least differences in the number of amino acids in cytochrome *c* compared to humans? Which organism has the greatest difference?

3. Notice the difference in the number of amino acids in cytochrome *c* between each organism and humans. How might these differences explain the relatedness of each organism to humans?

Molecular Biology The study of fossils, comparative anatomy, and embryology provides support for Darwin's theory of evolution by natural selection. Molecular biology is the study of gene structure and function. Discoveries in molecular biology have confirmed and extended much of the data already collected about the theory of evolution. Recall that mutations in genes are the source of variations upon which natural selection acts. You can learn more about how molecular biology provides evidence for evolution by exploring the table and figure below.

Comparing sequences	All organisms on Earth have genes that are made of DNA and work in similar ways. This supports the theory that all organisms are related. Scientists can study relatedness of organisms by comparing genes and proteins among living species. For example, nearly all organisms contain a gene that codes for cytochrome *c*, a protein required for cellular respiration. Some species, such as humans and rhesus monkeys, have nearly identical cytochrome *c*. The more closely related two species are, the more similar their genes and proteins are.
Divergence	Scientists have found that some stretches of shared DNA mutate at regular, predictable rates. Scientists use this "molecular clock" to estimate at what time in the past living species diverged from common ancestors. For example, as shown in the figure, molecular data indicates that whales and porpoises are more closely related to hippopotamuses than they are to any other living species.

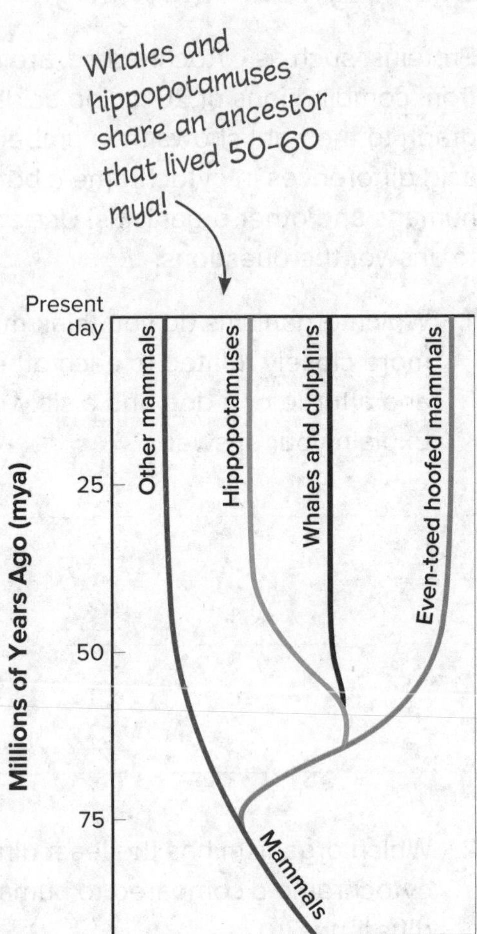

The Study of Evolution Today The theory of evolution by natural selection is the cornerstone of modern biology. Since Darwin published his theory, scientists have confirmed, refined, and extended Darwin's work. They have observed natural selection in hundreds of living species. Their studies of fossils, anatomy, embryology, and molecular biology have all provided evidence of relatedness among living and extinct species.

COLLECT EVIDENCE

How would you use molecular biology to help you understand the evolutionary relationship between bats and birds? Record your evidence (C) in the chart at the beginning of the lesson.

A Closer Look: From Dinosaur to Bird

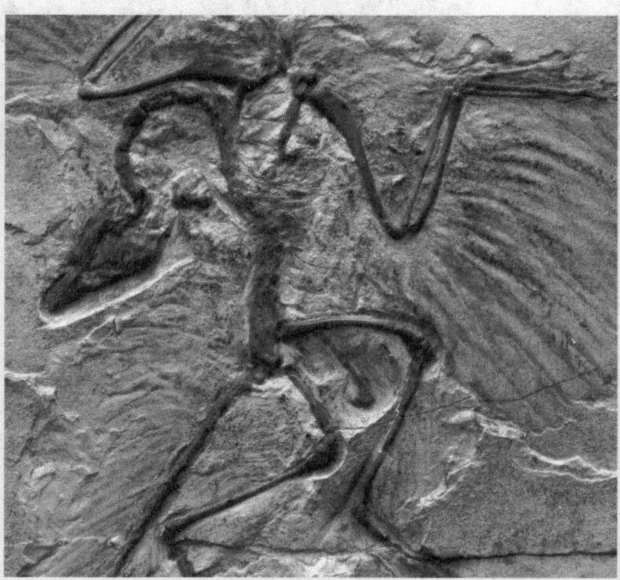

Scientists generally agree that dinosaurs were a type of reptile and that birds descended from reptiles. Scientists don't agree, however, on how closely birds are related to dinosaurs.

To see how closely related birds might be to dinosaurs, scientists compare and contrast living birds, such as chickens, with fossils of primitive species that might be related to birds. Using computer programs, they look for matches in at least 80 physical traits of modern birds. These traits include the skull, teeth, neck, pelvis, tail, shoulder, bones, feet, ankles, and stance.

Scientists have found several species of dinosaurs that had feathers. Others have bones that are similar to modern birds but unlike any other living animal. Some dinosaurs also had wrists that could bend in a flapping motion, like a wing, and toes that were arranged so they could grasp branches.

Researchers have recently extracted DNA and proteins from Mesozoic dinosaur fossils. This has allowed scientists to discover molecular evidence linking birds to dinosaurs.

It's Your Turn

Present With a partner, choose an organism and research its evolutionary history. Analyze and interpret data in the fossil record to determine similarities and differences between the organism and its fossil ancestors. Present its history in a blogpost by using narrative techniques, emphasizing relevant evidence and focused points.

LESSON 2

Review

Summarize It!

1. **Classify** Examine the list of structures below. Sort them into lists according to whether they are vestigial, homologous, or analogous features using this space. Then construct a three-page brochure that illustrates each set of structures and that explains what they are and why you have categorized them as you have. You may draw the structures or find pictures to cut out.

your nose	**airplane wing**	**human tailbone**
bat wing	**a pig's snout**	**butterfly wing**
whale pelvic bone	**an elephant's trunk**	**penguin wing**

Three-Dimensional Thinking

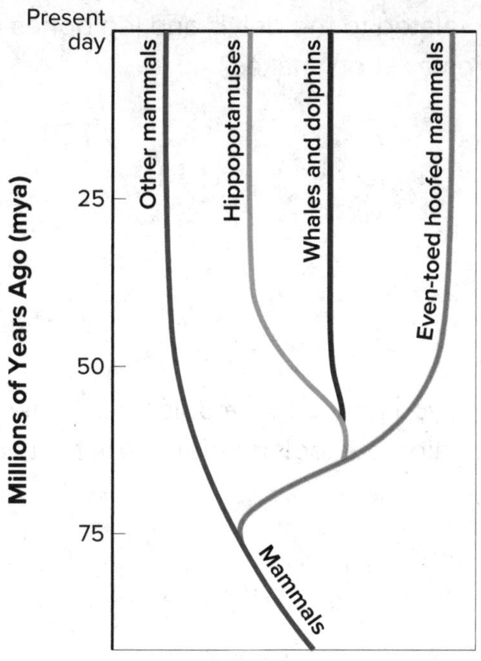

2. The image above shows that even-toed hoofed mammals and other mammals shared a common ancestor. When did this ancestor live?

A 25–35 million years ago

B 50–60 million years ago

C 60–75 million years ago

D 75 million years ago

---

3. Which developmental structure among vertebrates is evidence that they share a common ancestor?

A analogous structures

B pharyngeal pouches

C variation rates

D vestigial structures

Real-World Connection

4. Infer At the zoo you may see a boa or a python. If you look closely, you will see two small pelvic spurs, which are related to the pelvis and leg bones of other animals. Why do these structures exist on snakes?

5. Explain While watching a nature show, you notice that a dolphin's flipper and a shark's fin perform the same function but look different. What is this an example of? Explain your answer.

Still have questions?
Go online to check your understanding about biological evidence of evolution.

REVISIT SCIENCE PROBES

Do you still agree with the person you chose at the beginning of the lesson? Return to the Science Probe at the beginning of the lesson. Explain why you agree or disagree with that person now.

EXPLAIN THE PHENOMENON

Revisit your claim about how evolutionary relationships can be inferred. Review the evidence you collected. Explain how your evidence supports your claim.

PLAN AND PRESENT

STEM Module Project
Science Challenge

Now that you've learned about biological evidence of evolution, go back to your Module Project to plan your exhibit. You'll want to make connections between whales and other present-day organism to explain why whales have pelvic bones.

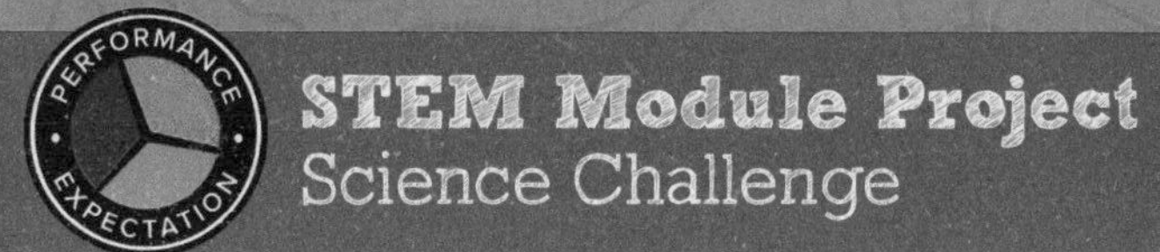

STEM Module Project
Science Challenge

It's All Relative

The director of a natural history museum wants to develop a series of exhibits that show how modern organisms are related to ancestral fossil organisms and to other modern organisms.

He asks you to prepare an exhibit about one specific modern organism and its relationships to fossilized organisms and modern organisms. Along with your exhibit, the director has asked you to prepare a written explanation for museum visitors. You will need to explain how the fossil and biological evidence can be used to make inferences about evolutionary relationships.

Planning After Lesson 1

How will you incorporate information about fossils and the fossil record in your museum exhibit?

Planning After Lesson 1, continued

In the space below, make a diagram of five undisturbed layers of sedimentary rock. Label your diagram using the terms "oldest" and "youngest" and use an arrow to show the pattern of relative age of the layers.

Explain how a diagram, similar to the one you prepared in the space above, could be used in your museum exhibit.

Planning After Lesson 2

Research to gather information about anatomical similarities between your organism and other living organisms. Specifically describe how this information could be used in your exhibit.

How will you incorporate information about embryology in your exhibit? Be sure to analyze and compare pictures of embryos across multiple species, including the species you have chosen.

Is information about the molecular biology of your organism as compared to other organisms available? If so, explain how this will be included in your project.

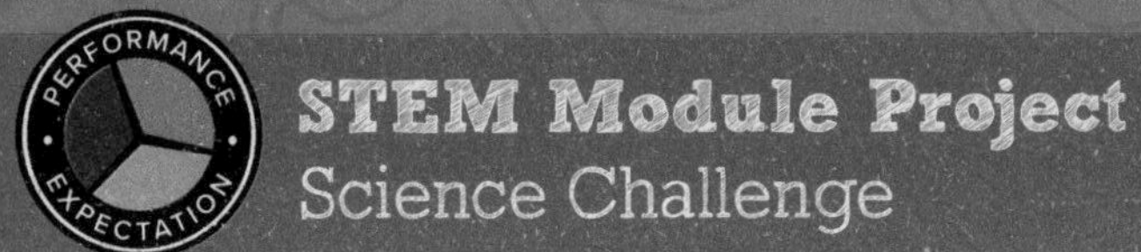

Research

Choose a modern organism to research and record your choice here:

Develop a research plan. The following may help guide you as you create this plan. Use the next page to take notes as you carry out your research.

Research the following to explore evolutionary relationships between this organism and extinct fossilized organisms:

- When does the fossil record indicate that the ancestors of this organism first appeared on Earth?
- Are there species in the fossil record that have similar structures?
- What does the evolutionary history of this organism look like based on the fossil record?
- What anatomical structures changed over time? Which structures stayed the same?

Research the following to explore evolutionary relationships between this organism and other modern organisms:

- What evidence from comparative anatomy provides clues about relationships between this organism and other modern organisms?
- Create a slide show on the developmental biology of your organism. Trade with a partner and examine their slide show, comparing it to what you have read in the text.
- Identify other lines of evidence can be used to determine evolutionary relationships between your organism and other modern organisms. What information from these lines of evidence will you include? What evidence will you include?

Use your Science Notebook to record your research.

Design Your Exhibit

On a poster board, design an exhibit that will effectively share your information with museum visitors. Sketch your exhibit, and label each visual element or type of information you will include. Include elements that will make your display visually appealing and interesting to museum visitors.

Construct Your Explanation

Apply the scientific ideas and information you have researched to construct an explanation. Your explanation should use evidence and reasoning to describe the information in your exhibit. As you work, consider the following points about your explanation.

- How does the explanation add information to the exhibit?

- Does it describe inferences that can be made about the evolutionary relationships among organisms based on similarities and differences in their anatomy and other lines of evidence? Explain.

- Does it identify patterns in the fossil record that are related to your topic?

Construct your explanation below.

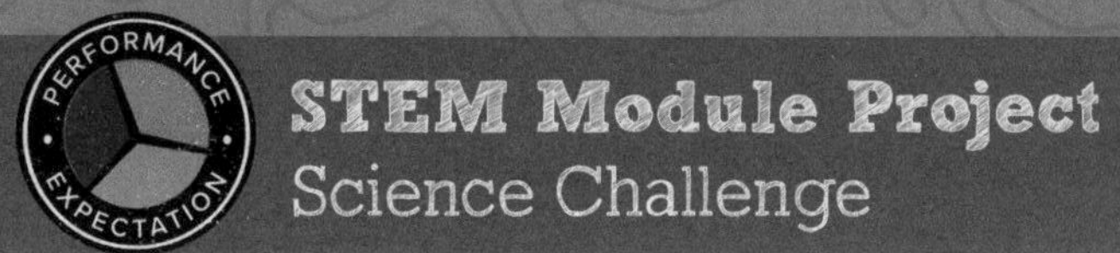

Create Your Presentation

Complete and finalize your presentation.

Present your exhibit and explanation to an audience of museum directors and museum visitors.

Go back to your research plan. What one thing might you do differently if you were asked to make an additional exhibit about a different organism?

Identify one strength and one weakness of your exhibit and explanation.

Do the exhibit and explanation you prepared help you better understand how pelvic bones in a whale can be used as evidence for evolution? Explain.

Congratulations! You've completed the Science Challenge requirements.

Module Wrap-Up

REVISIT THE PHENOMENON

Using the concepts that you have learned throughout this module, explain why a whale has pelvic bones even though it does not have legs.

INQUIRY

What are one or two questions you still have about the phenomenon?

Choose the question that interests you the most. Plan and conduct an investigation to answer this question.

Glossary

Multilingual Glossary

A science multilingual glossary is available on the science website. The glossary includes the following languages.

Arabic	Hmong	Tagalog
Bengali	Korean	Urdu
Chinese	Portuguese	Vietnamese
English	Russian	
Haitian Creole	Spanish	

Cómo usar el glosario en español:

1. Busca el término en inglés que desees encontrar.
2. El término en español, junto con la definición, se encuentran en la columna de la derecha.

Pronunciation Key

Use the following key to help you sound out words in the glossary.

a	back (BAK)	**Ew**	food (FEWD)
ay	day (DAY)	**yoo**	pure (PYOOR)
ah	father (FAH thur)	**yew**	few (FYEW)
ow	flower (FLOW ur)	**uh**	comma (CAH muh)
ar	car (CAR)	**u (+ con)**	rub (RUB)
E	less (LES)	**sh**	shelf (SHELF)
ee	leaf (LEEF)	**ch**	nature (NAY chur)
ih	trip (TRIHP)	**g**	gift (GIHFT)
i (i + com + e)	idea (i DEE uh)	**J**	gem (JEM)
oh	go (GOH)	**ing**	sing (SING)
aw	soft (SAWFT)	**zh**	vision (VIH zhun)
or	orbit (OR buht)	**k**	cake (KAYK)
oy	coin (COYN)	**s**	seed, cent (SEED)
oo	foot (FOOT)	**z**	zone, raise (ZOHN)

English — A — Español

absolute age/comparative anatomy

edad absoluta/anatomía comparativa

absolute age: the numerical age, in years, of a rock or object.

adaptation (a dap TAY shun): an inherited trait that increases an organism's chance of surviving and reproducing in a particular environment.

analogous (uh NAH luh gus) structures: body parts that perform a similar function but differ in structure.

edad absoluta: edad numérica, en años, de una roca o de un objeto.

adaptación: rasgo heredado que aumenta la oportunidad de un organismo de sobrevivir y reproducirse en su medioambiente.

estructuras análogas: partes del cuerpo que ejecutan una función similar pero tienen una estructura distinta.

C

camouflage (KAM uh flahj): an adaptation that enables a species to blend in with its environment.

comparative anatomy: the study of similarities and differences among structures of living species.

camuflaje: adaptación que permite a las especies mezclarse con su medioambiente.

anatomía comparativa: estudio de las similitudes y diferencias entre las estructuras de las especies vivas.

correlation: a method used by geologists to fill in the missing gaps in an area's rock record by matching rocks and fossils from separate locations.

correlación: método utilizado por los geólogos para completar vacios en un área de registro de rocas, comparando rocas y fósiles de lugares distanciados.

cross-cutting relationships: the principle that if one geologic feature cuts across another feature, the feature that it cuts across is older.

principio de las relaciones de corte: principio que dice que si un cuerpo de roca corta a otro debe ser más joven que el cuerpo de roca cortado.

D

DNA: the abbreviation for deoxyribonucleic (dee AHK sih ri boh noo klee ihk) acid, an organism's genetic material.

ADN: abreviatura para ácido desoxirribonucleico, material genético de un organismo.

E

embryology (em bree AH luh jee): the science of the development of embryos from fertilization to birth.

embriología: ciencia que trata el desarrollo de embriones desde la fertilización hasta el nacimiento.

extinction (ihk STINGK shun): event that occurs when the last individual organism of a species dies.

extinción: evento que ocurre cuando el último organismo individual de una especie muere.

F

fossil: the preserved remains or evidence of past living organisms.

fósil: restos conservados o evidencia de organismos vivos del pasado.

G

genetic engineering: the biological and chemical methods that change the arrangement of DNA that makes up a gene.

ingeniería genética: métodos biológicos y químicos que cambian la disposción del ADN que consiste en un gen.

geologic time scale: a chart that divides Earth's history into different time units based on changes in the rocks and fossils.

escala de tiempo geológico: tabla que divide la historia de la Tierra en diferentes unidades de tiempo, basado en los cambios en las rocas y fósiles.

H

homologous (huh MAH luh gus) structures: body parts of organisms that are similar in structure and position but different in function.

estructuras homólogas: partes del cuerpo de los organismos que son similares en estructura y posición pero diferentes en función.

I

inclusion: a piece of an older rock that becomes a part of a new rock.

inclusión: pedazo de una roca antigua que se convierte en parte de una roca nueva.

index fossil: a fossil representative of a species that existed on Earth for a short length of time, was abundant, and inhabited many locations.

fósil índice: fósil representativo de una especie que existió en la Tierra por un período de tiempo corto, ésta era abundante y habitaba en varios lugares.

K

key bed: a rock or sediment layer with distinctive characteristics that make it easily identifiable in correlation.

capa de sedimentos: capa de rocas o sedimentos con características que son identificables fácilmente en corelación.

L

lateral continuity: principle that sediments are deposited in large, continuous sheets in all lateral directions.

continuidad lateral: principio que dice que los sedimentos son despositados en capas grandes y continuas en direcciones laterales.

M

mass extinction: the extinction of many species on Earth within a short period of time.

extinción en masa: extinción de muchas especies en la Tierra dentro de un período de tiempo corto.

mimicry (MIH mih kree): an adaptation in which one species looks like another species.

mimetismo: una adaptación en el cual una especie se parece a otra especie.

mutation (myew TAY shun): a permanent change in the sequence of DNA, or the nucleotides, in a gene or a chromosome.

mutación: cambio permanente en la secuencia de ADN, de los nucleótidos, en un gen o en un cromosoma.

N

natural selection: the process by which organisms with variations that help them survive in their environment live longer, compete better, and reproduce more than those that do not have the variations.

selección natural: proceso por el cual los organismos con variaciones que las ayudan a sobrevivir en sus medioambientes viven más, compiten mejor y se reproducen más que aquellas que no tienen esas variaciones.

nucleotide (NEW klee uh tide): a molecule made of a nitrogen base, a sugar, and a phosphate group.

nucelótido: molécula constituida de una base de nitrógeno, azúcar y un grupo de fosfato.

O

original horizontality: principle that most rock-forming materials are deposited in horizontal layers.

principio de la horizontalidad original: principio que mantiene que la mayoria de materiales que forman las rocas son depositados en capas horizontales.

R

relative age: the age of rocks and geologic features compared with other nearby rocks and features.

edad relativa: edad de las rocas y de las características geológicas comparada con otras rocas cercanas y sus características.

replication: the process of copying a DNA molecule to make another DNA molecule.

replicación: proceso por el cual se copia una molécula de ADN para hacer otra molécula de ADN.

RNA: ribonucleic acid, a type of nucleic acid that carries the code for making proteins from the nucleus to the cytoplasm.

ARN: ácido ribonucleico, un tipo de ácido nucléico que contiene el código para hacer proteínas del núcleo para el citoplasma.

S

selective breeding: the selection and breeding of organisms for desired traits.

cría selectiva: proceso de cría de organismos para características deseadas.

superposition: the principle that in undisturbed rock layers, the oldest rocks are on the bottom.

superposición: principio que establece que en las capas de rocas inalteradas, la rocas más viejas se encuentran en la parte inferior.

T

transcription: the process of making mRNA from DNA.

transcripción: proceso por el cual se hace mARN de ADN.

transitional fossils: represent the intermediate evolutionary forms of life in transition from one type to another, or a common ancestor of these types.

translation: the process of making a protein from RNA.

formas transicionales: representan las formas intermedias evolutivas de vida en transición de un tipo a otro, o un antepasado común de especies diferentes.

traslación: proceso por el cual se hacen proteínas a partir de ARN.

U

unconformity: a surface where rock has eroded away, producing a break, or gap, in the rock record.

uniformitarianism: a principle stating that geologic processes that occur today are similar to those that occurred in the past.

discontinuidad: superficie donde la roca se ha erosionado, produciendo un vacío en el registro geológico sedimentario.

uniformimsmo: principio que establece que los procesos geológicos que ocurren actualmente son similares a aquellos que ocurrieron en el pasado.

V

variation (ver ee AY shun): a slight difference in an inherited trait among individual members of a species.

vestigial (veh STIH jee ul) structure: body part that has lost its original function through evolution.

variación: ligera diferencia en un rasgo hereditario entre los miembros individuales de una especie.

estructura vestigial: parte del cuerpo que a través de la evolución perdió la función original.

Index

Italic numbers = illustration/photo
Bold numbers = vocabulary term
lab = indicates entry is used in a lab
inv = indicates entry is used in an investigation
smp = indicates entry is used in a STEM Module Project
enc = indicates entry is used in an Encounter the Phenomenon
scc = indicates entry is used in a STEM Career Connection

M

N

O

P

R

S

T

U

V

W